Pollutant Dispersion in Built Environment

Tingzhen Ming · Chong Peng
Tingrui Gong · Zhengtong Li

Pollutant Dispersion in Built Environment

ZHEJIANG UNIVERSITY PRESS
浙江大学出版社

Springer

Tingzhen Ming
School of Civil Engineering and Architecture
Wuhan University of Technology
Wuhan, Hubei
China

Tingrui Gong
School of Energy and Power Engineering
Huazhong University of Science and Technology
Wuhan, Hubei
China

Chong Peng
School of Architecture and Urban Planning
Huazhong University of Science and Technology
Wuhan, Hubei
China

Zhengtong Li
School of Energy and Power Engineering
Huazhong University of Science and Technology
Wuhan, Hubei
China

The print edition is not for sale in China Mainland. Customers from China Mainland please order the print book from: Zhejiang University Press.

ISBN 978-981-10-3820-4 ISBN 978-981-10-3821-1 (eBook)
DOI 10.1007/978-981-10-3821-1

Jointly published with Zhejiang University Press, Hangzhou, China

Library of Congress Control Number: 2017931570

Printed on acid-free paper

This Springer imprint is published by Springer Nature
The registered company is Springer Nature Singapore Pte Ltd.
The registered company address is: 152 Beach Road, #21-01/04 Gateway East, Singapore 189721, Singapore

Preface

Since the severe haze and fog enveloped almost all the metropolises of China in the end of 2012, this situation has been becoming worse and worse. Presently, the haze and fog has become a common phenomenon which can be seen from East-North China through Northern China, Eastern China till Southern China, covering 1–2 million km^2. The haze and fog are mainly particulate matter less than 2.5 μm in the atmosphere mainly coming from the emissions of power plants, industrial factories, traffic and transportations, and residential sectors. Haze has adverse effects on human health, especially the early life of children and pregnant mothers. Further, haze, rich in toxic and hazardous substances, can directly enter the human body through the respiratory system and adhere to the upper and lower respiratory tract and lungs, which ultimately causes respiratory and cardiovascular diseases. Thereby, a further investigation on pollutant dispersion in outdoor built environment of cities, especially metropolises with densely population, plays an important role in improving the residents' thermal comfort and possibly provides the researchers and decision makers with a good guidance for architecture design, environmental assessment, energy application, urban planning, and so on.

In this book we are going to reveal the basic mechanisms of fluid flow, heat transfer, especially the transport of pollutants in outdoor built environment. We hope this book can be a good guidance for developers who are interested in building a healthy living environment.

In Chap. 1, we will present a brief introduction of the background of pollutant and haze in China, and we will also introduce the recent research development of how to test and analyze the pollutant dispersion in built environment during the past 20 years.

In Chap. 2, a comprehensive mathematical model including the mass, momentum, and energy equations to describe the fluid flow and heat transfer characteristics, species transport equation to describe the pollutant transport mechanism, and RNG k-ε turbulence model to analyze the turbulent flow has been put forward. The numerical simulation results are compared and verified with the existing experimental results. The effects of ambient crosswind velocity and direction, pollutant emission source intensity, the street canyon aspect ratio (building height versus street width) on the pollutant dispersion are analyzed.

Chapter 3, a numerical simulation was conducted using PHOENICS so as to assess the wind thermal environment of an urban area according to its current city structure, architectural composition, materials of ground surface, distribution of pollution source, environmental conditions, and solar radiation. The thermal environment and pollutant transport mechanism were analyzed, so were the problems and influential characteristics of current city layout.

In Chap. 4, we extended the heat transfer field synergy principle to mass transfer field synergy. With the comparison between temperature gradient and concentration gradient, we managed to analyze the synergy angle of velocity with concentration gradient, based upon which, we also revealed the relationship between synergy angle and pollutant concentration. Further, via the mass entransy dissipation derived from the heat transfer function, we analyzed the relationship between that and pollutant concentration.

In Chap. 5, a novel solar chimney with an inverted U-type cooling tower and a water spraying system (SCIUCTWSS) was proposed to mitigate the urban air pollution. In this system, an inverted U-type cooling tower was used to take place of the traditional chimney erected in the center of the collector; a water spraying system was installed at the turning point of the inverted U-type cooling tower to enhance the driving force; a filtration screen is placed near the entrance of the collector to filter out PM2.5 and large particulate matter from the airflow. The clean air out of the system's outlet can immediately improve the air quality in the spectrum of human activity. A mathematical model to describe the fluid flow, heat transfer of the system was further developed. Influence of injected water from the water spraying system on the pressure, velocity, temperature, and air density distributions were analyzed.

This book is supported by the Important Project of National Natural Science Foundation of China (No. 41590844), the Key program of National Natural Science Foundation of China (No. 51538004), the National Support Program Sub-topics of the 12th Five-Year Plan (No. 2011BAJ07B01-1), the National Natural Science Foundation of China (51478199), and the ESI Discipline Promotion Foundation of WUT (No.35400664).

Addition to the four authors of Professor Chong Peng, Mr. Tingrui Gong, and Mr. Zhengtong Li, several contributors have helped the authors to create this book. Mr. Jie Xu helped with the numerical simulations and draft preparation. Dr. Liquan Xu, Mr. Cunjin Cai, Wei Yang, Hongwen Dou, Weijie Fang, Jiayue Huang, Wenjie Deng, Yongjia Wu, Zuyuan Guo, Miss.Yingzi Lin, Ting Li, and Chen Zhang made contributions to data collection, figures, and tables treatment. They also did lots of work on tailoring contents, translating, and polishing this book.

Wuhan, China
October 2016

Tingzhen Ming

Contents

Chapter 1
Introduction

1.1 Pollutant Dispersion and Haze in Built Environment

There are several problems which arouse worldwide attention: energy crisis, environmental pollution, and water scarcity. Currently, about 90% of Chinese cities suffer from the severe haze phenomenon which is mainly caused by the emissions of fossil-fired power stations, the exhaust of vehicles, and the dust from numerous construction sites. Airborne aerosols, such as sulfate, nitrate, ammonium, particulate organic matter, black carbon, and other chemical species, can scatter and absorb the incident light and therefore lead to atmosphere opacity and horizontal visibility degrade. When the horizontal visibility is equal to or less than 10 km and atmospheric relative humidity is equal to or less than 90%, this phenomenon is called atmospheric haze [1]. The basic state policy of urbanization in China helps an increasing number of the rural population live in buildings with better working and living conditions. However, a significant increase of energy consumption during this urbanization process also causes a great pressure on the Government of P.R. China as the haze and fog phenomenon is becoming worse and worse. Haze has adverse effects on human health, especially the early life of children and pregnant mothers [2–7]. Haze, rich in toxic and hazardous substances, can directly enter the human body through the respiratory system and adhere to the upper and lower respiratory tract and lungs, which ultimately causes respiratory and cardiovascular diseases [8].

In high-density megacities, such as Beijing, Shanghai, Hong Kong, etc., air pollution has higher impact on the public health than those of lower population density. Presently, various adverse impacts of air pollution and haze on human activities have been witnessed, which has caused unaccountable loss and damages. Apart from higher pollution emissions due to human activities in densely populated street canyons, stagnated airflow due to closely packed tall buildings means lower dispersion potential. High-density urban morphologies need to be carefully designed to lessen the ill effects of high-density urban living [9]. In addition, the

T. Ming et al., *Pollutant Dispersion in Built Environment*,
DOI 10.1007/978-981-10-3821-1_1

research data showed that, in 2010, the residential sector accounted for around 18% of total energy consumption in China, but contributed 10, 50, and 69% of anthropogenic SO_2, black carbon, and organic carbon emissions, respectively [10].

1.2 Literature Review

The research on air pollution and it transport mechanism in built environment has attract worldwide attention. The research group led by Prof. Zhaolin Gu presented excellent work in the fields of transportation mechanism of particles such as sand and dust in the environment [11–21], the effect of green plants on air quality [22], the effect of air quality on urban design [23, 24], and numerical simulation method of particles in various ambience [16, 19, 24–27]. The research group led by Prof. Yinping Zhang comprehensively considered seriously the volatile organic compounds (VOCs) emissions, their adverse effect on residents, the measurement methods, and the treatment method in indoor environment [7, 28–41]. The research group led by Prof. Qihong Deng mainly considered the effects of various particles on children and pregnant mothers' health [3–7, 42–53]. In addition, quite a lot of researchers focused on the pollutant dispersion in street canyons as transport emissions are regarded as a main contribution in air pollution of metropolis.

1.2.1 Experimental Methods and Results

In 2002, Sagrado et al. [54] experimentally studied the pollutant dispersion in a two-dimensional street canyon in the L-2B wind tunnel at von Karman Institute in which they mainly considered the influence of the height of the downstream building. In their research, the pollutant source was situated in the middle of the street. The concentration measurements were performed by means of light scattering technique and the velocity field was measured with particle image velocimetry (PIV). Two different sets of configurations were defined: open country and non-isolated street canyon. In the open country cases, only the street canyon was studied without any other buildings around. In the non-isolated street canyon a third building was placed upstream of the street to study its influence on the flow and dispersion in the street. In both sets of configurations three different heights of the downstream building were investigated. From the measurements it was found that increasing the height of the downstream building decreases the pollutant concentration in the street. In both cases, open country and non-isolated street, the flow is drastically separated and accelerated when impacting against the first building placed upstream.

Rotach et al. [55] reported the first results of an urban tracer experiment being realized in the framework of the "Basel Urban Boundary Layer Experiment" in an area with abundant information on turbulence and flow conditions available. In this experimental, release height and the height of concentration samples were close to roof level; the meteorological conditions were mainly convective; the mechanical turbulence was considered substantial due to the rough character of the underlying surface. The concentration distribution was found to be essentially Gaussian in the horizontal plane and some commonly used methods to estimate the plume widths in applied dispersion models were compared to the observations. From measurements at one site downwind of the source it was found that for a near-roof level source, only an insignificant vertical gradient in tracer concentration was present within a street canyon.

Ning et al. [56] experimentally and numerically studied the dispersion characteristics of vehicular exhaust plume at an idle condition in an idealized and simplified environment. The gaseous and particulate concentrations in the exhaust plume of three idling motor vehicles were measured in an isolated environment under calm weather conditions. They found that despite the difference in the initial concentrations, the pollutants decayed exponentially in all directions. Later, the effects of the initial emission concentration, exit velocity, exit direction, and crosswind intensity were investigated parametrically. The results indicated that the initial pollutant concentration increases the local concentrations but the pattern of dispersion remains the same; the exit velocity increases the momentum of the jet, resulting in a deeper penetration downstream; the exit angle has a stronger influence on pollutant dispersion than both initial pollutant concentration and exit velocity.

Said et al. [57] explored the underlying organized motions, i.e. coherent structures, within the near-wake region of a turbulent round jet discharged perpendicularly from a chimney into a cross-flow with the aid of PIV technique. This cross-flow was found to be quite complex owing to its three-dimensional nature and the interactions between several flow regions. They examined the wake regions of the chimney and plume at levels near the top of the chimney and discussed the complex geometry of these structures in the wake of the plume as well as their interaction with the plume as it bends over after emission. Later, they described the Kelvin–Helmholtz vortex structures, the downwash phenomena, and the effect of the height of the chimney.

Chung and Chung [58] estimated the vehicle emissions by measuring the traffic flow rates and modeling the vehicle emissions as banded line sources along the tunnel floor. The effects of fan ventilation and piston effect of moving vehicles on the airflow and pollutant dilution were examined numerically. The results revealed that a peak velocity exists near the tunnel floor due to the piston effect of vehicles. The cross-sectional concentrations of air pollutants are nonuniformly distributed and concentrations rise with downstream distance. The piston effect of vehicles can alone provide 25–34% dilution of air pollutants in the tunnel, compounded 43–70% dilution effect according to the ventilation condition.

Gromke and Ruck [59] performed flow field and concentration measurements in an idealized model of an urban street canyon with one row of trees arranged along the center axis. The model was set up in an atmospheric boundary layer wind tunnel and the approach flow was directed perpendicular to the street axis. A line source embedded in the bottom of the street was used to release tracer gas for the simulation of traffic exhaust emissions. Trees with spherical crowns were modeled and positioned inside the street canyon, varying crown diameter, crown permeability, trunk height, and tree spacing. Traffic-induced turbulence was simulated by rotating belts with thin plates. Concentrations were measured at the facades of the street canyon. For small tree crowns, only little changes in concentration were measured, however, increasing crown diameters led to increasing concentrations at the leeward street canyon wall associated with a reduction of local concentrations at the windward wall. For some cases, a variation of trunk height led to a modification of the concentration pattern on the walls. Increasing the tree spacing resulted in a noticeable concentration decrease.

Jiang et al. [60] conducted an investigation on the flow and pollutant dispersion in an urban street canyon with parallelepiped buildings and towers by numerical simulation method and water tank experiments. Two types of canyon geometry were tested. The studies indicated that in a step-up notch canyon (higher buildings on the downstream side of the canyon), the height and shape of the upstream lower buildings plays an important role in flow pattern and pollutant dispersion, while in a step-down notch canyon (lower buildings on the downstream side), the downstream lower buildings have little influence; the substitution of tall towers for parallelepiped buildings on one side of the canyon may enhance the street ventilation and decrease the pollutant concentration emitted by motor vehicles.

Ma et al. [61] conducted experimental measurements and numerical simulations of air distribution and microorganism pollutant distribution in the auditorium and game area in a gymnasium and put forward the methods to pollutant dispersion control and to improve the indoor air quality. The results showed that the fungi and bacteria concentration levels are less than the magnitude of 10^3 CFU (colony-forming units) which meets the requirements of indoor air quality standard. The numerical simulation results quantitatively agreed with the experimental data while some differences between theoretical data and experimental data existed in air distributions. They also found that people number in gymnasium plays an important role in affecting indoor air quality and the environmental parameters attained the standard.

Garbero et al. [62] investigated the dispersion of a passive scalar within an idealized urban district made up of a building-like obstacle array. They focused on a street network in which the lateral dimension of the buildings exceeds the street width, a geometry representative of many European cities. They performed a series of wind tunnel experiments to investigate the effect of different geometries and wind directions on the pollutant dispersion process. Concentration measurements of a passive tracer have enabled us to infer the main features charactering its dispersion within the street network. We describe this by focusing on the roles of different transfer processes. These are the channeling of the tracer along the street axes, the mixing at street intersections, and the mass exchange between the streets and the overlying atmospheric flow. Our experiments provide evidence of the dependence of these processes on the geometrical properties of the array and the direction of the overlying atmospheric flow.

1.2.2 Numerical Methods and Results

Bottema [63] presented an analytical model for the evaluation of the roughness length and the displacement height of regular building groups. The results showed that apart from the frontal density (ratio of building frontal area and ground surface), the streamwise building length and the building layout pattern are important parameters. The method can be applied to the roughness mapping of towns, optimization of street design with respect to air ventilation, and optimization, and tuning of wind tunnel approach flow roughness.

Gotting et al. [64] advanced a three-dimensional microscale numerical model MIMO to calculate the dispersion of a passive pollutant in the vicinity of a U-shaped building. In their research, three different locations of a non-buoyant source and six different wind directions were examined. The calculated concentrations were compared with concentrations measured in a boundary layer wind tunnel, which showed a very good agreement between numerical and wind tunnel results.

Gross [65] studied the airflow and pollutant dispersion in street canyons by using the microscale model ASMUS-F. The results demonstrated a strong dependence on car-induced turbulence, while the shape of roofs and trees along the street modify the situation only locally. Kumar et al. [66] developed an analytical puff model based on parameterization schemes of atmospheric boundary layer theory to analyzed the pollutant dispersion problems.

Li and Stathopoulos [67] evaluated the air pollutant distribution around a building by solving the concentration equation based on the previously simulated wind flow field. They applied the hybrid scheme for the evaluation of pollutant concentration around a rectangular building for two different sources under conditions of neutral atmospheric stratification. The simulation results were compared to the available experimental data from previous studies in boundary layer wind tunnels. It showed that the numerical data agreed well far downwind of the building but agree less satisfactorily close to the wall and within the wake zone.

Massons et al. [68] presented a numerical study of nonreactive pollutant dispersion in sea breeze conditions. Sea breeze circulation is investigated using a 3-D mesoscale meteorological model. Simulation was conducted for the area of Tarragona (Spain) which has an important petrochemical industry in the coastal region and complex terrain. Results from the meteorological model were used as input to a Lagrangian particle model in order to analyze the pollutant dispersion of an elevated plume emitting near the shoreline. The simulation was performed for 24 h and an analysis of the meteorological and concentration fields was undertaken for this time period. The results are compared with measured surface data. Good correlation exists between observed and simulated conditions indicating that the coupling of the meteorological and particle models provides a good tool for analyzing air pollution in complex situations.

Selvam [69] A computer model based on the k-epsilon turbulence model is used to compute flow around a building. Using this simulation flow-held pollutant concentrations around the building are computed using upwind and streamline upwind Petro–Galerkine (SUPG) finite element procedure. In this study the source is kept adjacent to the bottom (position 1) and top (position 2) of the leeward center wall. The computed concentrations using upwind and SUPG procedure are compared with wind-tunnel measurements. The SUPG procedure computes the trend of the concentration much closer to the wind tunnel measurement than the upwind procedure.

Chabni et al. [70] studied the flow dynamics and the concentration dispersion in urban street canyons by using large eddy simulation method to study the influence of the instantaneous flow on the concentration field. In addition, the LES method was particularly used to investigate the two-dimensional characteristics of the concentration field established for flows perpendicular to the street with emissions released near the floor of the street by motor vehicles. The results showed that, for a street with $D/H = 2$, where D is the width of the street and H is the height of the building, the average statistical solution presented two counter-rotating whirls. The concentration in the leeward region of the canyon, compared with the windward one, is significant when the wind is perpendicular to the axis of the canyon.

Hassan and Crowther [71] carried out a two-dimensional steady-state numerical simulation for a typical street canyon ventilated by a crosswind. They used the PHOENICS package from CHAM to analyze the air flow above and within the street canyon. The k-epsilon turbulence model was used for turbulence modeling and pollutant sources were added at ground level over the road but not over the pavements. The results for the air flow showed the formation of a longitudinal vortex within the street canyon. Pollutant concentrations were predicted with the highest values occurring at the leeward walls of the upwind buildings, and the lowest values on the windward walls of the downwind buildings. The accuracy of these simulations was examined by comparing the predicted results with field observations.

Li and Stathopoulos [72] presented the numerical simulation of air pollutant dispersion around cubic and rectangular buildings by using the standard k-epsilon turbulence model. Results were compared with experimental data and past numerical simulation results. The influence of convergence criteria to the numerical solution was investigated and the discretization error was estimated by using two different grid systems. Specific threshold values for convergence criteria were recommended.

Chan et al. [73] developed a two-dimensional numerical model based on Reynolds-averaged Navier–Stokes equations coupled with a series of standard, Renormalization Group (RNG) and realizable kappa-epsilon turbulence models to simulate the fluid flow development and pollutant dispersion within an isolated street canyon using the FLUENT code. They found that in their study, the RNG k-epsilon turbulence model was found to be the most optimum turbulence model coupled with the two-dimensional street canyon model developed; both the calculated and measured dimensionless pollutant concentrations were shown to be less dependent on the variation of wind speed and source strength conditions for the studied street canyon aspect ratio of the $D/H = 1$ case. However, the street canyon configuration has significant influence on the pollutant dispersion. The wider street and lower height of the buildings are favorable to pollutant dilution within the street canyon. The fluid flow development has demonstrated that the rotative vortex or vortices generated within the urban street canyon can transport the pollutants from a line source to the wall surfaces of the buildings.

Baker et al. [74] numerically investigated turbulent flow in an urban street canyon to the dispersion and transport of reactive pollutants (NO, NO_2 and O_3) by using large eddy simulation model. They studied a real scale street canyon with an aspect ratio of one under the working conditions being a neutrally stratified atmosphere and initial wind perpendicular to the street axis. A background concentration of ozone was specified and the gases were allowed to mix and react. Results indicated that spatial variation of pollutants within the canyon was calculated to be significant.

Xie et al. [75] carried out a numerical analysis to reveal the impact of solar radiation on pollutant dispersion in different urban street layouts using CFD technique. It was indicated that the street geometry has a strong influence on the wind field and pollution dispersion patterns.

Tsai et al. [76] conducted a three-dimensional numerical simulation to analyze the effects of excess ground and building temperatures on airflow and dispersion of pollutants in an urban street canyon. In this research, vehicle emissions were estimated from the measured traffic flow rates and modeled as banded line sources, with a street length and bandwidths equal to typical vehicle widths. Both measurements and simulations revealed that pollutant concentrations typically follow the traffic flow rate, decline as the height increases, and become higher on the leeward side than on the windward side.

Blocken et al. [77] conducted steady-state RANS CFD simulations of pollutant dispersion in the neutrally stable atmospheric boundary layer with the commercial code Fluent 6.1 for three case studies: plume dispersion from an isolated stack, low-momentum exhaust from a rooftop vent on an isolated cubic building model, and high-momentum exhaust from a rooftop stack oil a low-rise rectangular building with several rooftop structures. The results were compared with the Gaussian model, the semiempirical ASHRAE model, and wind tunnel and full-scale measurements. It was shown that in all three cases and with all turbulence models tested, the lateral plume spread is significantly underestimated. It was suggested that transient simulations might be required to achieve more accurate results. The numerical results were quite sensitive to the value of the turbulent Schmidt number.

Wang et al. [78] investigated the likely route of dispersion of airborne pollutants and pathogenic agents in the re-entrance space of a high-rise residential (HRR) building in Hong Kong. In their research, a 1:150 scaled 30-storey HRR building model was tested in a wind tunnel with simulated atmospheric boundary layer flow. Tracer gas, propane released from the lower, middle and upper parts of the building, was measured by flame ionization detection which was used to simulate the air pollutant dispersion within the re-entrance space. The results showed that the airborne pollutants or pathogenic agents could be dispersed mainly along the vertical routes, posing a potential risk to the adjacent neighbors immediately above the source location. Wind direction could also affect the dispersion.

Zhang et al. [79] studied the air flow and pollutant dispersion characteristics in an urban street canyon by proposing a new scheme for real-time boundary conditions. The induced effect of time series of ambient wind conditions on the flow structures inside and above the street canyon was investigated. The flow showed an obvious intermittent feature in the street canyon and the flapping of the shear layer forms near the roof layer under real-time wind conditions, resulting in the expansion or compression of the air mass in the canyon. The simulations of pollutant dispersion showed that the pollutants inside and above the street canyon were transported by different dispersion mechanisms, relying on the time series of air flow structures. Large scale air movements in the processes of the air mass expansion or compression in the canyon exhibited obvious effects on pollutant dispersion.

Uneven building layouts and nonuniform street canyons are common in actual urban morphology. Gu et al. [80] studied the effects of various building layouts on air flow in nonuniform street canyons; carried out Large eddy Simulation (LES) under four cases with parameter change of the occupying ratio of high buildings in the computational domain and their bilateral allocation as well as the combinations of step-up and/or step-down notches. The results indicated that, at the pedestrian level, the concentrations of simulated pollutants in the nonuniform street canyons are lower than those in the uniform one, suggesting that uneven building layouts are capable of improving the dispersion of pollutants in urban area; the separation of step-up and step-down notches in non-uniform street canyons might be a good choice for uneven building layout arrangements from the point of view of pollutant dispersion and human health.

Liu et al. [81] numerically investigated the wind field and pollutant dispersion in an urban area with crowded buildings and heavy traffic by LES method with Lagrangian dynamic model to close the subgrid scale stress. The wind speed and temperature fields are computed from 7 am on September 6 2005 to 7 am of the next day and the pollutant concentration field is computed from 7 am to 6 pm on September 6 2005. The wind field and pollutant concentration distribution were demonstrated in the site by proposed combined model and they were compared with those computed by drag element model in whole computational domain. The results showed that the combined model is an appropriate method for predicting wind field and pollutant dispersion in the crowded building area.

Liu et al. [82] developed a backward time modeling of diffusion convection pollutant dispersions with quasi-reversibility method. The procedure is applied to the backward time identification of the contaminant release history and source location in a three-dimensional slot ventilated building enclosure. Spatial distributions of pollutant concentrations were known in priori. The effects of supplying air velocity, pollutant source location, pollutant diffusivity property, and pollutant release time on the accuracy of the pollutant dispersion history recovery were investigated. Numerical results demonstrated that the accuracy of the pollutant dispersion history recovery can be enhanced with different approaches and measures, including the promotion of room ventilation rate, the shrinkage of distance between the pollutant source and supplying airport, and the reduction of pollutant diffusivity.

Mechanical ventilation poses some challenges for control of pollutants in a fully enclosed dental clinic. Zhang et al. [83] built an advanced turbulence model including large eddy simulation for the flow field and the discrete phase modeling for pollutant tracks to study the effect of natural ventilation on pollutant dispersion in a dental clinic setting. Three basic ventilation paths were identified, the single narrow path, narrow path and dispersive path. The results showed that the single narrow path had the highest efficiency, with an escape time of about 1/30 and 1/100 of the narrow and dispersive paths, respectively. Despite the position of the pollutant source and facilities such as bulkheads, escape time was significantly reduced when the ventilation flow rate was increased under the single narrow and dispersive paths.

Cui et al. [84] numerically and experimentally studied the Reynolds number independence (Re-independence) of turbulent flow and pollutant dispersion in urban areas. The concept of numerical wind tunnel (NWT) was proposed and validated by prototype wind tunnel experimental measurements. A new physical quantity: the ratio of relative change (RRC) was proposed to investigate the Re-independence quantitatively. For the given street canyon, numerically predicted variations of RRC versus building Reynolds number (Re-H) from three k-epsilon turbulence models agreed well with each other, and the variation trend showed that there exist two flow regimes in the range of Re-H studied: flow in low Re-H region is strongly affected by Re-H, while that in higher Re-H region is weakly affected by Re-H, giving a strong support to the concept of Re-independence of turbulent flow.

Miao et al. [85] used the Open Source Field Operation and Manipulation (OpenFOAM) software package to numerically study the fluid flow and traffic exhaust dispersion in urban street canyons with different configurations to find out the urban planning strategies to ease the air pollution. The patterns of flow and dispersion within three different kinds of street canyon configuration under the perpendicular approaching flow were also studied. The result showed that the width and height of building can dramatically affect the pollution level inside the street canyon. As the width or height of building increases, the pollution at the pedestrian level increases.

Yuan et al. [9] addressed the knowledge gap between planning and design principles and air pollution dispersion potentials in high-density cities. In this study, the air ventilation assessment for projects in high-density Hong Kong was advanced to include air pollutant dispersion issues. Urban scale parametric studies were conducted to clarify the effects of urban permeability and building geometries on air pollution dispersion, for both the outdoor pedestrian environment and the indoor environment in the roadside buildings. A real urban case study was finally conducted to demonstrate that the suggested design principles from the parametric study are feasible in the practical high-density urban design.

Huang et al. [86, 87] presented a series of two-dimensional numerical simulations to describe the fluid flow and pollutant dispersion in an urban street canyon by using the FLUENT code. They considered the effects of the strength and position of pollutant sources, the shapes of height of upstream roofs on pollutant dispersion within an urban street canyon. The results indicated that the four parameters above significantly affect the fluid flow and pollutant dispersion.

Zhang et al. [88] proposed a numerical wind tunnel model to compute the pollutant diffusion around a typical Hong Kong high-rise building model (at a linear scale of 1:30) and found that the pathways of pollutant migration for windward and leeward pollutant emission were different. The simulated results of haze–fog (HF) studies confirmed that the pathway of pollutant migration is dominated by wind structure interaction and buoyancy effect only plays a minor role in the dispersion process.

Cui et al. [89] built a multiscale physical model to study the basic phenomena of air flow, heat and mass transfer passing through different scale urban areas. Wind tunnel measurements were conducted in TJ-1 and TJ-4 wind tunnels with ground heating to model solar radiation and SF6 released from upwind building rooftop. Particle image velocimetry (PIV), T-type thermocouples and SF6 quantitative detector were used to measure flow fields, temperature and pollutant distributions. Thermal effects on flow structures were modeled for Richardson numbers (Ri) from 0 to 4.77. Measurement results showed that for the investigation of mixed convection in the street canyon by wind tunnel, if the value of Reynolds number is in the range of Re-independence determined by forced boundary layer flow, the only criterion for flow pattern and temperature distribution of the external flow is the Richardson number.

Mei et al. [90] numerically investigated the pedestrian level pollutant transport in street canyons with multiple aspect ratios (H/W), considering various unstable thermal stratification scenarios and plain surrounding, solve the non-isothermal turbulent wind flow, temperature field, and pollutant spread within and above the street canyons by the realizable k-epsilon turbulence model along with the enhanced wall treatment. The numerical simulation results indicated that: (1) one-vortex flow regime is observed for shallow canyons with $H/W = 0.5$, whereas multi-vortex flow regime is observed for deep canyons with $H/W = 2.0$; (2) both one-vortex and multi-vortex regimes could be observed for the street canyons with $H/W = 1.0$, where the secondary vortex could be initiated by the flow separation and intensified by unstable thermal stratification.

Mei et al. [91] presented a numerical simulation on the dust particles transportation within urban street canyons using Reynolds-averaged Navier–Stokes (RANS) method, and obtained air solid two-phase flow fields on a regional scale. In this study, four typical street building models, namely, low-rise buildings, step-up building arrangements, step-down building arrangements, and high-rise buildings were analyzed where the particle volume fraction distribution reflected the basic properties of particle transportation in the canyons. To evaluate the dust particle transport performance in the models, three indices, namely particle transport efficiency, suspension fraction and suspension density, were defined. The results showed that the variation of the streamwise air velocity, vertical velocity, and fluctuating velocity on the roof of canyons along the x direction were separately examined; the fluctuating velocity was the dominant mechanism of the particle suspension in the canyons.

1.3 Contents of This Book

In this Chapter, a very brief review on the pollutant dispersion in street canyon has been presented. In Chap. 2, numerical simulation was performed to analyze the fluid flow, heat transfer, and pollutant dispersion of several street canyons. Mathematical model including the mass equation, momentum equation, energy equation, turbulence equation, and the species (pollutants) transport equation was advanced. Then the effects of pollutant source intensity, ambient crosswind velocity and direction, and ratio of street height versus street width on the pollutant dispersion in the street canyons have been analyzed.

In Chap. 3, taking the street buildings of a county as an example, three-dimensional geometrical and mathematical models were advanced, the temperature, velocity, and pollutant concentration distributions were presented. In Chap. 4, the Filed synergy analysis, which is generally used in heat transfer enhancement, was introduced to analyze the pollutant transport mechanism in outdoor built environment. In chap. 5, a novel solar chimney with an inverted U-type cooling tower and a water spraying system (SCIUCTWSS) was proposed to mitigate the urban air pollution. A mathematical model to describe the fluid flow, heat transfer of the system was further developed. Influence of injected water from the water spraying system on the pressure, velocity, temperature, and air density distributions were analyzed.

References

1. D. Wu, X. Bi, X. Deng, F. Li, H. Tan, G. Liao, J. Huang, Effects of atmospheric haze on the deterioration of visibility over the Pearl River Delta, Acta Meteorologica Sinica, 64 (2006) 510–517.
2. C. Lu, Q.H. Deng, Y.G. Li, J. Sundell, D. Norback, Outdoor air pollution, meteorological conditions and indoor factors in dwellings in relation to sick building syndrome (SBS) among adults in China, Science of the Total Environment, 560 (2016) 186–196.
3. Q.H. Deng, C. Lu, C.Y. Ou, L. Chen, H. Yuan, Preconceptional, prenatal and postnatal exposure to outdoor and indoor environmental factors on allergic diseases/symptoms in preschool children, Chemosphere, 152 (2016) 459–467.
4. Q.H. Deng, C. Lu, Y.G. Li, J. Sundell, D. Norback, Exposure to outdoor air pollution during trimesters of pregnancy and childhood asthma, allergic rhinitis, and eczema, Environmental Research, 150 (2016) 119–127.

5. Y. Lu, X. Liu, Q.H. Deng, Y.Z. Duan, H.J. Dai, Y. Li, T. Xiao, X.P. Ning, J.L. Fan, L. Zhou, X.H. Li, H. Zhong, H. Yuan, Continuous Lead Exposure Increases Blood Pressure but Does Not Alter Kidney Function in Adults 20–44 Years of Age in a Lead-Polluted Region of China, Kidney Blood Pressure Res., 40(3) (2015) 207–214.
6. Q.H. Deng, C. Lu, C.Y. Ou, W.W. Liu, Effects of early life exposure to outdoor air pollution and indoor renovation on childhood asthma in China, Building and Environment, 93 (2015) 84–91.
7. Q.H. Deng, C. Lu, D. Norback, C.G. Bornehag, Y.P. Zhang, W.W. Liu, H. Yuan, J. Sundell, Early life exposure to ambient air pollution and childhood asthma in China, Environmental Research, 143 (2015) 83–92.
8. X.G. Liu, J. Li, Y. Qu, T. Han, L. Hou, J. Gu, C. Chen, Y. Yang, X. Liu, T. Yang, Y. Zhang, H. Tian, M. Hu, Formation and evolution mechanism of regional haze: a case study in the megacity Beijing, China, Atmospheric Chemistry and Physics, 13 (2013) 4501–4514.
9. C. Yuan, E. Ng, L.K. Norford, Improving air quality in high-density cities by understanding the relationship between air pollutant dispersion and urban morphologies, Building and Environment, 71 (2014) 245–258.
10. J. Liu, D.L. Mauzerall, Q. Chena, Q. Zhang, Y. Song, W. Peng, Z. Klimont, X. Qiu, S. Zhang, M. Hu, W. Lin, K.R. Smith, T. Zhu, Air pollutant emissions from Chinese households A major and underappreciated ambient pollution source, proceedings of the national academy of science of the United States of America, 113(28) (2016) 7756–7761.
11. Z.L. Gu, L.Y. Lu, L. Ai, Impact energy loss from particle surface roughness in particulate systems, Epl, 87(1) (2009) 4.
12. Z.L. Gu, J.W. Su, J.Y. Jiao, X.Y. Xu, Simulation of micro-behaviors including nucleation, growth, and aggregation in particle system, Sci. China Ser. B-Chem., 52(2) (2009) 241–248.
13. L.Y. Lu, Z.L. Gu, K.B. Lei, An inter-particle contact area and time restoration for softening treatment in thermal discrete element modeling, Epl, 87(4) (2009) 4.
14. L.Y. Lu, Z.L. Gu, K.B. Lei, S. Wang, K. Kase, An efficient algorithm for detecting particle contact in non-uniform size particulate system, Particuology, 8(2) (2010) 127–132.
15. J.W. Su, Z.L. Gu, X.Y. Xu, Discrete element simulation of particle flow in arbitrarily complex geometries, Chem. Eng. Sci., 66(23) (2011) 6069–6088.
16. W. Wei, Z.L. Gu, S. Wang, T. Fukuda, K. Kase, J. Ju, Y. Yamagata, Y. Tajima, Numerical simulation of nanoparticle pattern fabricated by electrostatic spray deposition, Particuology, 11(1) (2013) 20–24.
17. J.W. Su, Z.L. Gu, M.X. Zhang, X.Y. Xu, An improved version of RIGID for discrete element simulation of particle flows with arbitrarily complex geometries, Powder Technol., 253 (2014) 393–405.
18. J.W. Su, Z.L. Gu, C.G. Chen, X.Y. Xu, A two-layer mesh method for discrete element simulation of gas-particle systems with arbitrarily polyhedral mesh, Int. J. Numer. Methods Eng., 103(10) (2015) 759–780.
19. J.W. Su, C.Q. Huang, Z.L. Gu, C.G. Chen, X.Y. Xu, An Efficient RIGID Algorithm and Its Application to the Simulation of Particle Transport in Porous Medium, Transp. Porous Media, 114(1) (2016) 99–131.
20. L.Y. Zhang, Z.L. Gu, C. Yu, Y.W. Zhang, Y. Cheng, Surface charges on aerosol particles - Accelerating particle growth rate and atmospheric pollution, Indoor and Built Environment, 25(3) (2016) 437–440.
21. M.X. Zhang, X.L. Luo, T.Y. Li, L.Y. Zhang, X.Z. Meng, K. Kase, S. Wada, C.W. Yu, Z.L. Gu, From Dust Devil to Sustainable Swirling Wind Energy, Sci Rep, 5 (2015) 5.
22. Z.L. Gu, Y.W. Zhang, K.B. Lei, Large eddy simulation of flow in a street canyon with tree planting under various atmospheric instability conditions, Sci. China-Technol. Sci., 53(7) (2010) 1928–1937.
23. Y.W. Zhang, Z.L. Gu, Air quality by urban design, Nat. Geosci., 6(7) (2013) 506–506.
24. Y.W. Zhang, Z.L. Gu, Z.S. Wang, Y. Cheng, F.S.C. Lee, Advances in the Fine Scale Simulation of Urban Wind Environment, Indoor and Built Environment, 22(1) (2013) 332–336.

25. W. Wei, Z.L. Gu, S. Wang, Y.W. Zhang, K.B. Lei, K. Kase, Numerical simulation of the cone-jet formation and current generation in electrostatic spray-modeling as regards space charged droplet effect, J. Micromech. Microeng., 23(1) (2013) 11.
26. Z.L. Gu, J.Y. Jiao, Y.W. Zhang, J.W. Su, Large eddy simulation using a dynamic mixing length subgrid-scale model, Int. J. Numer. Methods Fluids, 69(9) (2012) 1457–1472.
27. Z.L. Gu, Y.Z. Zhao, Y. Li, Y.Z. Yu, X. Feng, Numerical simulation of dust lifting within dust devils - Simulation of an intense vortex, J. Atmos. Sci., 63(10) (2006) 2630–2641.
28. J.Y. Xiong, J.P. Cao, Y.P. Zhang, Early stage C-history method: Rapid and accurate determination of the key SVOC emission or sorption parameters of indoor materials, Building and Environment, 95 (2016) 314–321.
29. Y.X. Wu, S.S. Cox, Y. Xu, Y.R. Liang, D.Y. Won, X.Y. Liu, P.A. Clausen, L. Rosell, J.L. Benning, Y.P. Zhang, J.C. Little, A reference method for measuring emissions of SVOCs in small chambers, Building and Environment, 95 (2016) 126–132.
30. C. Liu, Y.P. Zhang, Characterizing the equilibrium relationship between DEHP in PVC flooring and air using a closed-chamber SPME method, Building and Environment, 95 (2016) 283–290.
31. Z.M. Bu, Y.P. Zhang, D. Mmereki, W. Yu, B.Z. Li, Indoor phthalate concentration in residential apartments in Chongqing, China: Implications for preschool children's exposure and risk assessment, Atmos. Environ., 127 (2016) 34–45.
32. Z.M. Bu, L.F. Wang, L.B. Weschler, B.Z. Li, J. Sundell, Y.P. Zhang, Associations between perceptions of odors and dryness and children's asthma and allergies: A cross-sectional study of home environment in Baotou, Building and Environment, 106 (2016) 167–174.
33. W.J. Wei, J.Y. Xiong, W.P. Zhao, Y.P. Zhang, A framework and experimental study of an improved VOC/formaldehyde emission reference for environmental chamber tests, Atmos. Environ., 82 (2014) 327–334.
34. C. Liu, Y.P. Zhang, C.J. Weschler, The impact of mass transfer limitations on size distributions of particle associated SVOCs in outdoor and indoor environments, Science of the Total Environment, 497 (2014) 401–411.
35. Z.J. Du, J.H. Mo, Y.P. Zhang, Q.J. Xu, Benzene, toluene and xylenes in newly renovated homes and associated health risk in Guangzhou, China, Building and Environment, 72 (2014) 75–81.
36. Z.J. Du, J.H. Mo, Y.P. Zhang, Risk assessment of population inhalation exposure to volatile organic compounds and carbonyls in urban China, Environment International, 73 (2014) 33–45.
37. W.J. Wei, J.Y. Xiong, Y.P. Zhang, Influence of Precision of Emission Characteristic Parameters on Model Prediction Error of VOCs/Formaldehyde from Dry Building Material, PLoS One, 8(12) (2013) 8.
38. F. Qu, L.B. Weschler, J. Sundell, Y.P. Zhang, Increasing prevalence of asthma and allergy in Beijing pre-school children: Is exclusive breastfeeding for more than 6 months protective?, Chin. Sci. Bull., 58(34) (2013) 4190–4202.
39. W.W. Liu, Y.P. Zhang, Y. Yao, Labeling of volatile organic compounds emissions from Chinese furniture: Consideration and practice, Chin. Sci. Bull., 58(28–29) (2013) 3499–3506.
40. C. Liu, Z. Liu, J.C. Little, Y.P. Zhang, Convenient, Rapid and Accurate Measurement of SVOC Emission Characteristics in Experimental Chambers, PLoS One, 8(8) (2013) 9.
41. S.D. Huang, J.Y. Xiong, Y.P. Zhang, A rapid and accurate method, ventilated chamber C-history method, of measuring the emission characteristic parameters of formaldehyde/VOCs in building materials, Journal of Hazardous Materials, 261 (2013) 542–549.
42. Y. Yang, Q.H. Deng, Numerical Modelling to Evaluate the Disinfection Efficacy of Multiple Upper-Room Ultaviolet Germicidal Fixtures System, in: Y. Sun, J. Pei (Eds.) 9th International Symposium on Heating, Ventilation and Air Conditioning, Elsevier Science Bv, Amsterdam, 2015, pp. 1657–1664.

43. W.W. Liu, Z.W. Lian, Q.H. Deng, Use of mean skin temperature in evaluation of individual thermal comfort for a person in a sleeping posture under steady thermal environment, Indoor and Built Environment, 24(4) (2015) 489–499.
44. Q.H. Deng, C. Lu, C.W.F. Yu, Characterizing ambient concentration of PM10 in urban environment of central south China, Indoor and Built Environment, 24(3) (2015) 324–339.
45. H.J. Dai, Z.J. Huang, Q.H. Deng, Y. Li, T. Xiao, X.P. Ning, Y. Lu, H. Yuan, The Effects of Lead Exposure on Serum Uric Acid and Hyperuricemia in Chinese Adults: A Cross-Sectional Study, Int. J. Environ. Res. Public Health, 12(8) (2015) 9672–9682.
46. C. Lu, Q.H. Deng, C.W.F. Yu, J. Sundell, C.Y. Ou, Effects of ambient air pollution on the prevalence of pneumonia in children: Implication for National Ambient Air Quality Standards in China, Indoor and Built Environment, 23(2) (2014) 259–269.
47. Z.J. Huang, Y.Q. Zhou, H. Yuan, Y.Z. Duan, X.H. Tang, Q.H. Deng, PM2.5 components in association with emergency department visits for hypertension-related cardiovascular disease, J. Am. Coll. Cardiol., 64(16) (2014) C93–C93.
48. Z.J. Huang, Y.Q. Zhou, Y. Lu, H. Yuan, X.H. Tang, C. Lu, Q.H. Deng, Y.Z. Duan, Outdoor air pollution and hospital emergency room visit for stroke in a Chinese southern city, J. Am Coll. Cardiol., 64(16) (2014) C103–C103.
49. Y.P. Zhang, B.Z. Li, C. Huang, X. Yang, H. Qian, Q.H. Deng, Z.H. Zhao, A.G. Li, J.N. Zhao, X. Zhang, F. Qu, Y. Hu, Q. Yang, J. Wang, M. Zhang, F. Wang, X.H. Zheng, C. Lu, Z.J. Liu, Y.X. Sun, J.H. Mo, Y.L. Zhao, W. Liu, T.T. Wang, D. Norback, C.G. Bornehag, J. Sundell, Ten cities cross-sectional questionnaire survey of children asthma and other allergies in China, Chin. Sci. Bull., 58(34) (2013) 4182–4189.
50. C. Lu, Q.H. Deng, C.Y. Ou, W.W. Liu, J. Sundell, Effects of ambient air pollution on allergic rhinitis among preschool children in Changsha, China, Chin. Sci. Bull., 58(34) (2013) 4252–4258.
51. W.W. Liu, Q.H. Deng, W.W. Ma, H.F. Hao, J. Zhao, Feedback from human adaptive behavior to neutral temperature in naturally ventilated buildings: Physical and psychological paths, Building and Environment, 67 (2013) 240–249.
52. W.W. Liu, Y. Zheng, Q.H. Deng, L. Yang, Human thermal adaptive behaviour in naturally ventilated offices for different outdoor air temperatures: A case study in Changsha China, Building and Environment, 50 (2012) 76–89.
53. W.W. Liu, Z.W. Lian, Q.H. Deng, Evaluation of Human Thermal Comfort using the Mean Skin Temperature, Southeast Univ Press, Nanjing, 2009.
54. A.P.G. Sagrado, J. van Beeck, P. Rambaud, D. Olivari, Numerical and experimental modelling of pollutant dispersion in a street canyon, J. Wind Eng. Ind. Aerodyn., 90(4–5) (2002) 321–339.
55. M.W. Rotach, S.E. Gryning, E. Batchvarova, A. Christen, R. Vogt, Pollutant dispersion close to an urban surface - the BUBBLE tracer experiment, Meteorology and Atmospheric Physics, 87(1–3) (2004) 39–56.
56. Z. Ning, C.S. Cheung, Y. Lu, M.A. Liu, W.T. Hung, Experimental and numerical study of the dispersion of motor vehicle pollutants under idle condition, Atmos. Environ., 39(40) (2005) 7880–7893.
57. N.M. Said, H. Mhiri, G. Le Palec, P. Bournot, Experimental and numerical analysis of pollutant dispersion from a chimney, Atmos. Environ., 39(9) (2005) 1727–1738.
58. C.Y. Chung, P.L. Chung, A numerical and experimental study of pollutant dispersion in a traffic tunnel, Environ. Monit. Assess., 130(1–3) (2007) 289–299.
59. C. Gromke, B. Ruck, Influence of trees on the dispersion of pollutants in an urban street canyon - Experimental investigation of the flow and concentration field, Atmos. Environ., 41 (16) (2007) 3287–3302.
60. Y.J. Jiang, H.Z. Liu, J.G. Sang, B.Y. Zhang, Numerical and experimental studies on flow and pollutant dispersion in urban street canyons, Advances in Atmospheric Sciences, 24(1) (2007) 111–125.

61. F. Ma, L.R. Huo, H. Xie, H.F. Fan, Experimental and numerical simulation of air distribution and microorganism pollutant dispersion in gymnasium, Journal of Central South University of Technology, 16 (2009) 38–42.
62. V. Garbero, P. Salizzoni, L. Soulhac, Experimental Study of Pollutant Dispersion Within a Network of Streets, Boundary-Layer Meteorology, 136(3) (2010) 457–487.
63. M. Bottema, Urban roughness modelling in relation to pollutant dispersion, Atmos. Environ., 31(18) (1997) 3059–3075.
64. J. Gotting, C. Winkler, M. Rau, N. Moussiopoulos, G. Ernst, Dispersion of a passive pollutant in the vicinity of a U-shaped building, Int. J. Environ. Pollut., 8(3–6) (1997) 718–726.
65. G. Gross, ASMUS - A numerical model for simulations of wind and pollutant dispersion around individual buildings .2. Dispersion modelling and applications, Meteorol. Z., 6(3) (1997) 130–136.
66. P. Kumar, R.P. Mathur, P.K. Pande, P.N. Godbole, A puff algorithm for predicting pollutant dispersion based on ABL parameterization, J. Wind Eng. Ind. Aerodyn., 72(1–3) (1997) 117–125.
67. Y. Li, T. Stathopoulos, Numerical evaluation of wind-induced dispersion of pollutants around a building, J. Wind Eng. Ind. Aerodyn., 67–8 (1997) 757–766.
68. J. Massons, J. Camps, M.R. Soler, Modeling of pollutant dispersion in sea breeze conditions using a Lagrangian model, Theor. Appl. Climatol., 56(3–4) (1997) 255–266.
69. R.P. Selvam, Numerical simulation of pollutant dispersion around a building using FEM, J. Wind Eng. Ind. Aerodyn., 67–8 (1997) 805–814.
70. A. Chabni, P. Le Quere, C. Tenaud, H. Laatar, Modelling of pollutant dispersion in urban street canyons by means of a large-eddy simulation approach, Int. J. Veh. Des., 20(1–4) (1998) 88–95.
71. A.A. Hassan, J.M. Crowther, Modelling of fluid flow and pollutant dispersion in a street canyon, Environ. Monit. Assess., 52(1–2) (1998) 281–297.
72. Y. Li, T. Stathopoulos, Computational evaluation of pollutant dispersion around buildings: Estimation of numerical errors, J. Wind Eng. Ind. Aerodyn., 77–8 (1998) 619–630.
73. T.L. Chan, G. Dong, C.W. Leung, C.S. Cheung, W.T. Hung, Validation of a two-dimensional pollutant dispersion model in an isolated street canyon, Atmos. Environ., 36(5) (2002) 861–872.
74. J. Baker, H.L. Walker, X.M. Cai, A study of the dispersion and transport of reactive pollutants in and above street canyons - a large eddy simulation, Atmos. Environ., 38(39) (2004) 6883–6892.
75. X.M. Xie, Z. Huang, J.S. Wang, Z. Xie, The impact of solar radiation and street layout on pollutant dispersion in street canyon, Building and Environment, 40(2) (2005) 201–212.
76. M.Y. Tsai, K.S. Chen, C.H. Wu, Three-dimensional modeling dispersion in an urban street of air flow and pollutant canyon with thermal effects, Journal of the Air & Waste Management Association, 55(8) (2005) 1178–1189.
77. B. Blocken, T. Stathopoulos, P. Saathoff, X. Wang, Numerical evaluation of pollutant dispersion in the built environment: Comparisons between models and experiments, J. Wind Eng. Ind. Aerodyn., 96(10–11) (2008) 1817–1831.
78. J.H. Wang, J.L. Niu, X.P. Liu, C.W.F. Yu, Assessment of Pollutant Dispersion in the Re-entrance Space of a High-rise Residential Building, Using Wind Tunnel Simulations, Indoor and Built Environment, 19(6) (2010) 638–647.
79. Y.W. Zhang, Z.L. Gu, Y. Cheng, S.C. Lee, Effect of real-time boundary wind conditions on the air flow and pollutant dispersion in an urban street canyon-Large eddy simulations, Atmos. Environ., 45(20) (2011) 3352–3359.
80. Z.L. Gu, Y.W. Zhang, Y. Cheng, S.C. Lee, Effect of uneven building layout on air flow and pollutant dispersion in non-uniform street canyons, Building and Environment, 46(12) (2011) 2657–2665.
81. Y.S. Liu, G.X. Cui, Z.S. Wang, Z.S. Zhang, Large eddy simulation of wind field and pollutant dispersion in downtown Macao, Atmos. Environ., 45(17) (2011) 2849–2859.

82. D. Liu, F.Y. Zhao, H.Q. Wang, E. Rank, History source identification of airborne pollutant dispersions in a slot ventilated building enclosure, International Journal of Thermal Sciences, 64 (2013) 81–92.
83. W. Zhang, C.M. Mak, H.M. Wong, Pollutant dispersion in a natural ventilated dental clinic, Building Services Engineering Research & Technology, 34(3) (2013) 245–274.
84. P.Y. Cui, Z. Li, W.Q. Tao, Investigation of Re-independence of turbulent flow and pollutant dispersion in urban street canyon using numerical wind tunnel (NWT) models, International Journal of Heat and Mass Transfer, 79 (2014) 176–188.
85. Y.C. Miao, S.H. Liu, Y.J. Zheng, S. Wang, Y. Li, Numerical Study of Traffic Pollutant Dispersion within Different Street Canyon Configurations, Advances in Meteorology, (2014).
86. Y.D. Huang, W.R. He, C.N. Kim, Impacts of shape and height of upstream roof on airflow and pollutant dispersion inside an urban street canyon, Environmental Science and Pollution Research, 22(3) (2015) 2117–2137.
87. Y.D. Huang, X. Xu, Z.Y. Liu, C.N. Kim, Effects of Strength and Position of Pollutant Source on Pollutant Dispersion Within an Urban Street Canyon, Environmental Forensics, 16(2) (2015) 163–172.
88. Y. Zhang, K.C.S. Kwok, X.R. Liu, J.L. Niu, Characteristics of air pollutant dispersion around a high-rise building, Environmental Pollution, 204 (2015) 280–288.
89. P.Y. Cui, Z. Li, W.Q. Tao, Wind-tunnel measurements for thermal effects on the air flow and pollutant dispersion through different scale urban areas, Building and Environment, 97 (2016) 137–151.
90. S.J. Mei, C.W. Liu, D. Liu, F.Y. Zhao, H.Q. Wang, X.H. Li, Fluid mechanical dispersion of airborne pollutants inside urban street canyons subjecting to multi-component ventilation and unstable thermal stratifications, Science of the Total Environment, 565 (2016) 1102–1115.
91. D. Mei, Q. Deng, M. Wen, Z. Fang, Evaluating Dust Particle Transport Performance within Urban Street Canyons with Different Building Heights, Aerosol and Air Quality Research, 16 (2016) 1483–1496.

Chapter 2
Heat Transfer and Pollutant Dispersion in Street Canyons

2.1 Introduction

The rapid economic development with a dramatic growth of urbanization has induced many environmental issues. The emission of various pollutants into the atmospheric is one of the most severe pollutions, which directly associated people's health problems in urban air environment. Controlling air pollution remains a great challenge because of the diversity of sources and the complex interaction between atmospheric boundary layer flows and flow around building obstacles. Since vehicular exhausts are by far the major emission source of urban air pollution [1], it appears essential to seek solutions to mitigate traffic-related pollutants in urban environmental research.

With the pollutants are dispersed over a wide range of horizontal length scales, the wind field can be divided into the urban canopy and the urban boundary layers [2]. The urban canopy extends from the ground surface up to the level of the buildings, which involves the interaction of the plume and the flow field of the lower part of the urban boundary layer, may be perturbed by the geometry of the buildings and of the streets or other obstacles. In contrast, the urban boundary layers, which is above the buildings, has relatively small influence on a dispersion field due to its horizontal motion prevails over the vertical motion.

Street canyons are the basic geometric unit of street network in urban areas, which promotes the accumulation of traffic-induced pollution. To minimize the effect of traffic air pollution in built-up areas, it is necessary to model and accurately predict the flow and transport of pollutants in street canyons. Generally, street canyons are characterized by geometrical parameters such as the mean height of the buildings (H), the canyon width (B), and the canyon length (L). The pollutant dispersion process in street canyons is manyly influenced by ambient conditions (e.g., wind speed and direction), the geometric configurations, atmospheric stability, source strength of pollutants, and solar irradiation [3].

T. Ming et al., *Pollutant Dispersion in Built Environment*,
DOI 10.1007/978-981-10-3821-1_2

2.2 The Behavior of Pollutant Dispersion

In the last few decades, many experimental and numerical studies have been performed for the flow and the pollutants transport mechanisms in the street canyons.

Micro-scale Computational Fluid Dynamic (CFD) simulation as an emerging analysis method for pollutants dispersion, which is easily implemented various dispersion processes and able to reproduce the qualitative features of the flow field and pollutant concentration variance within the street canyons, gradually favored by researchers. Thus it together with wind tunnel and field measurements has become important methods in urban environmental research. Various investigations on the prediction of flow and pollutant dispersion within street canyons were performed using two-dimensional or three-dimensional steady-state Reynolds-averaged Navier–Stokes (RANS) equations coupled with the different turbulence models. By comparison, LES is able to capture the intermittent and unsteady fluctuations of the flow field, also comes with an expensive computational cost [4], although there is little difference between RANS and LES results for mean flow [5]. For two-dimensional (2D) street canyon models, the street aspect ratio which is defined as a ratio of the building height (H) to the width (B), is an important factor that determines flow regimes. Accordingly, these regimes can be categorized into the isolated roughness flow regime (IRF, $H/B < 0.3$), the wake interference flow regime (WIF, $0.3 < H/B < 0.67$), the skimming flow regime (SF, $0.67 < H/B < 1.67$) [2, 6, 7]. Besides, as the aspect ratio increases, there are two or more vortexes in relatively deep street canyons ($H/B > 1.67$), called the fourth flow regime of the multi-vortex regime by Refs. [8, 9]. For such 2D deep canyons, recirculation flows result in poor capacity of dispersing ground-level pollutants, which are mixed in canyons by vortices, thus the traffic-related pollutants are difficult to be removed out through deep canyons.

The Gaussian model of plume dispersion is widely used for pollutant concentration estimates [10–13], different flow regimes with corresponding resistances may be responsible for different pollutant concentration distributions in the urban boundary layer [14]. Sini et al. [6] developed a two-dimensional numerical model with the standard k–ε two-equation turbulence model to study the flows and vertical exchange of pollutants within the street canyons, the results showed that the number and arrangement of vortex structures within the street canyon can largely influence the vertical exchange rates. Moreover, it noted that differential heating may influence the vortex structures, from a one-vortex flow regime to multi-vortex flow regime. Chan et al. [15] performed a two-dimensional RANS numerical model coupled with standard, Renormalization Group (RNG), and realizable k–ε turbulence models to simulate the flow and pollutant dispersion within an isolated street canyon. It was found that the RNG k–ε turbulence model is the most optimum turbulence model in their research, and the dimensionless pollutant concentration is shown to be less dependent on the variation of wind speed and pollutant source strength for the street canyon aspect ratio $B/H = 1$, while the street canyon configuration has significant influence on it. The wider street, the lower height of canyon is, and the more pollutants being diluted will be. Cheng et al. [16] studied the ventilation behavior in different street canyons configurations, different vortexes such as clockwise vortex and anticlockwise vortex presented in upwind and downwind of the street canyon would lead to different levels of concentration of pollutants. A two-dimensional numerical model has been developed using the FLUENT code by Huang et al. [17], the flow field and pollutant dispersion within an urban street canyon with aspect ratio $B/H = 1$ are examined numerically considering five different shapes. Moreover, the impacts of elevated expressway on the wind flow and pollutant dispersion in the canyon are also investigated [18]. Miao et al. [19] studied the patterns of flow and dispersion within three different kinds of street canyon configuration under perpendicular approaching flow, it was found that the width and height of building can dramatically affect the pollution level within the street canyon, the nonuniform street canyon could provide better ventilation. Meanwhile, they indicated that it is not always a perpendicular flow that dominates the urban area, specific study is necessary for a particular area. Furthermore, we may notice this fact that most previous studies always considered the arranging direction of street is perpendicular to the wind direction. The incident wind direction is nearly 0° that can increase the flow capacity of pollutant dilution by mean flows and weaken turbulent mixing, which confirmed that is conductive to pollutant dispersion [20, 21].

As source strength also is a crucial parameter in determining the pollutant dispersion in the street canyon, Huang et al. [22] studied numerically the airflow and pollutant dispersion inside a street canyon by a two-dimensional numerical model using the FLUENT code, they found that the dimensionless pollutant concentrations were independent from the source strength with the aspect ratio $B/H = 1$. The position of pollutants sources also plays a very important role on pollutant distributions.

In order to investigate the effect of multiple aspect ratios (H/B) on the pedestrian level pollutant transport in street canyons, Mei et al. [23] investigated non-isothermal turbulent wind flow by solving the realizable k–ε turbulence model along with the enhanced wall treatment. The results show that one-vortex flow regimes are observed for shallow canyons with $H/B = 0.5$, whereas multi-vortex flow regimes are existed in the street canyons with $H/B = 2$. They indicated that the secondary vortex could be initiated by the flow separation and intensified by unstable thermal stratification. As a matter of fact, the temperature difference between the ground surface and inflow wind has significant influence on the flow field and pollutant dispersion. Mills [24] developed a computer simulation model to evaluate the urban canopy layer, which includes both buildings and street canyons. It found that changing the aspect ratio H/B can effectively alter the urban canopy layer energy fluxes and canyon air temperature. After that, Kim and Baik [3] performed a study of thermal effects on the flow and pollutant dispersion in street canyons by using a two-dimensional numerical model. It was shown that the dispersion of pollutant is quite dependent upon aspect ratio and heat source. Since the temperature difference plays an important role on the number and intensity of vortices in street canyons as well as the dispersion features, the building wall or street canyon bottom heating can significantly affect flow and dispersion. A photochemistry coupled CFD based numerical model has been applied to idealized street canyons of different aspect ratios from 0.5 to 8 with two different ambient wind speeds under different diurnal heating scenarios by Tong and Leung [25]. The results showed that different diurnal heating scenarios could substantially affect the reactive gases exchange between the street canyon and urban boundary layer. In addition, higher building aspect ratio and stronger ambient wind speed caused enhanced entrainment of O_3 concentrations into the street canyons along windward walls. Actually, the bottom heating leads to strong buoyancy forces as the air is heated by the wall surface when receiving direct solar radiation, which plays a significant role in determining the flow field and pollutant dispersion within street canyons [26–28].

Based on the above review, some features of two-dimensional (2D) pollutant dispersion within the street canyon can be identified by considering the configurations, ambient condition, source strength of pollutants and thermal effects. Indeed, pollutant dispersion is controlled by turbulent mixing and mean flow transport through the street canyon, but the realistic flow structure around a building is highly three-dimensional and complex when the wind approaches the building, which includes curved streamlines, velocity discontinuities, and nonhomogeneous turbulence disperse effluents. Since contaminant is caught in sidewall, rooftop, and some cavity recirculation zones around a group of buildings, thus it can be transported in the opposite direction of the prevalling wind due to eddies and mean flows induced by buildings [29, 30]. Meroney [31] indicated that the properties of a dispersion plume frequently depends upon the shape and intensity of motion within separated flow regions about the obstacle. Belcher also demonstrated that the dynamics of the shear layers and dividing streamlines play a decisive role on mean transport and turbulence mixing. In short, as Tominaga and Stathopoulos [5] suggested, it is necessary to reproduce three-dimensional flow to accurately predict the flow and pollutant dispersion around street canyons.

Li et al. [32, 33] experimentally and numerically investigated the isothermal turbulent flows and pollutant removal mechanism through high-rise urban areas with a parallel approaching wind. It was found that wind is quickly weakened due to the strong drag force produced by buildings. Strong shear layers exists near canopy roof levels which can produce 3D vortexes and considerable air exchanges across the canyon roofs and the interfaces of streets. The 3D canyon vortexes may be destroyed or deviated with building height variations, downward mean flow in front of taller buildings and upward flow behind taller buildings thus induced. In addition, the contribution of pollutant removal by mean flow and turbulent diffusion and ventilation capacity are examined. Yassin et al. [34] developed three-dimensional flow and dispersion of gaseous pollutants numerically investigated the impact of three cases of street configurations on pollution dispersion within urban street canyons. The results indicate that the street intersections are important factors on the flow field and pollutant dispersion in street canyons. The flow characteristics of different aspect ratios on the dispersion of reactive pollutants in street canyons were investigated using a coupled CFD-chemistry model by Park et al. [35]. In addition, small-scale models of Hong Kong were simulated via a 3D RNG k–ε turbulence model to investigate the influence of stratification on city ventilation by Yang and Li [36]. The results show that the impact of thermal stratification on airflow is minor when the wind speed is small, whereas the airflow in the street canyons is dependent on thermal stratification when the wind speed is small relative to buoyancy force, refer to Richardson numbers. Miao et al. [37] also found that the flow and turbulent fields within the street canyon under the more unstable Planetary Boundary Layer condition are stronger, resulted in less pollutants accumulated. Ming et al. [38, 39] numerically analyzed the wind and thermal environment by developing a mathematical model to describe the fluid flow and heat transfer characteristics of an old district in Wuhan, considered the ambient wind, solar radiation, natural convection, and radiation heat transfer. Some basic rules and optimizing local renewal strategies are compared to achieve favorable thermal comfort and flow ventilation effects.

From previous studies, the impact factors on pollutant dispersion process in street canyons can be attributed to the geometric configurations, wind speeds, source strength of pollutants and thermal effect. In this chapter, we develop two-dimensional numerical model and three-dimensional numerical model based on the RANS equations coupled with RNG turbulence model to simulate the pollutant dispersion within an isolated street canyon. Airborne pollutants (used CO) from vehicle exhausts are assumed as constant volumetric sources at pedestrian level. For two-dimensional model, the temperature at land surface is assumed constant and higher than that of incoming wind flow to simulate the thermal effect. For three-dimensional model, solar radiation will be imposed on the ground surface and all the building surfaces, varying with time, thus radiation model is added. The simulated data are validated against wind tunnel experimental results. Detailed analyses of street canyon configurations at different street widths, wind speeds, source strengths, and thermal effect on the flow field, temperature distributions, and pollutant dispersion within the street canyon are demonstrated.

2.3 Mathematical Model

The geometry models and boundary conditions in computation are schematized as Fig. 2.1. For two-dimensional model, five streets canyons are formed by six rows of buildings, as shown in Fig. 2.1a. For three-dimensional model, a street canyon is formed by two buildings, as shown in Fig. 2.1b. According to the AIJ guidelines for practical applications of CFD [40], the distance between the winds blowing boundary plane and the leading edge of the first building is set as 5H, where H is the building height. The lateral distance from boundary plane to building edge is also set as 5H for 3D model. The distance between outflow boundary and leeward edge of the last building is 15.5H. The upper boundary is set as 6H to the ground surface. The inlet wind speed was assumed to follow a power law profile:

$$U(z) = U_{\text{ref}}\left(\frac{z}{0.02}\right)^{0.2} \tag{2.1}$$

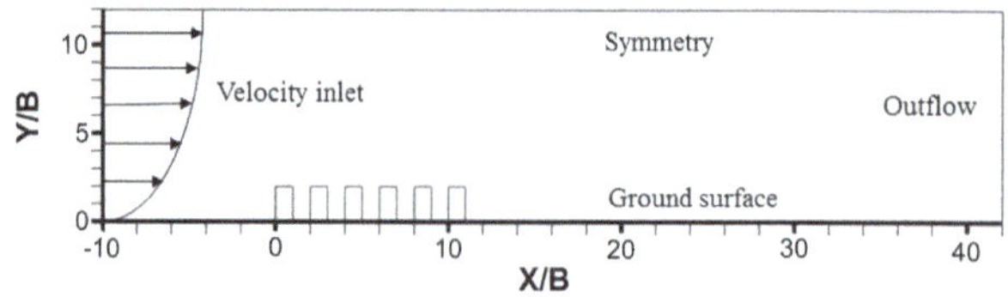

(a) 2D model

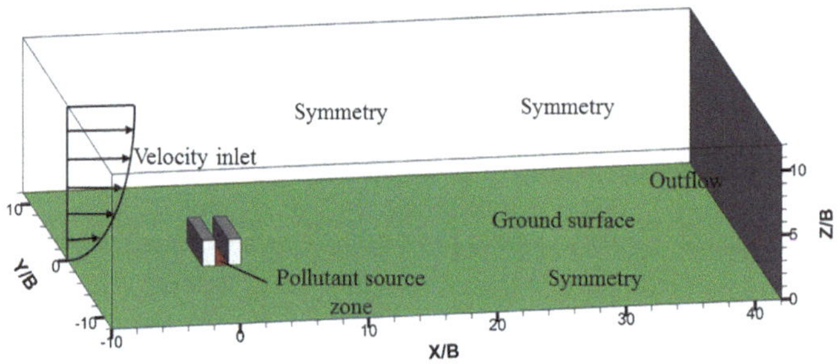

(b) 3D model

Fig. 2.1 Geometry models

Turbulent kinetic energy and dissipation rate at inlet boundary were specified as:

$$k = \frac{U_*^2}{\sqrt{C_\mu}} \tag{2.2}$$

$$\varepsilon = \frac{U_*^3}{\kappa z}, \tag{2.3}$$

where U_{ref} represents the velocity at a reference height. U_* represents the friction velocity. κ is the von Karman constant (0.4), and $C_\mu = 0.09$. At the upper boundary and at the lateral boundary (3D), symmetry boundary condition is applied, i.e., zero normal gradient. Free outflow boundary condition is applied at outlet. In 2D model, the temperature of the ground surface is assumed as constant, and the building surfaces are assumed as adiabatic. Whereas in 3D model, the solar radiation model is adopted, the heat transfer coefficient between the building surfaces and air is assumed as 25 W/(m^2 K). The pollutant release at pedestrian level (1–2 m) is represented by a source zone Q at the street canyon.

As mentioned, the flow field, if the perpendicular approaching wind is relative small, is considered to be purely natural convection induced by solar radiation which heats the ground surface and building walls and then results in the increase of air temperature. Richardson number is used to represent the importance of natural convection relative to the forced convection. However, the approaching wind plays a significant role in previous investigations as the buildings are all exposed to the ambient, which causes the system to be a combination of natural and forced convection. The natural flow should be vertically upward due to the effect of horizontal ground surface having high temperature; however, the airflow horizontally enhances this heat transfer effect, which puts the ground surface at a lower temperature level. The flow regimes can also be classified according to the ratio of Gr and Re^2 shown as:

$$\frac{Gr}{Re^2} = \frac{(g\beta\Delta TL^3/v^2)}{(U_{\text{ref}}L/v)^2} = \frac{g\beta\Delta TL}{U_{\text{ref}}^2}, \tag{2.4}$$

where Gr and Re are Grashof and Reynolds numbers, respectively; ΔT is the maximum temperature increase within the area; v, β, and L are, respectively, the air viscosity, thermal expansion coefficient, and the maximum building height; g is the acceleration of gravity. U_{ref} is the wind velocity at reference height in the area.

After the preliminary calculation, we found that Gr is higher than 10^{10} which is the upper limit transition value of natural convection from laminar to turbulent flow. Therefore, the turbulent mathematical model needs to be selected to describe fluid flow within the system. The RNG $k-\varepsilon$ turbulence model is chosen to evaluate numerical performance for the pollutant dispersion in an isolated street canyon. Also, the density variation in the whole computational model is so small. The density variation of the air is caused by temperature changes, rather than that of the pressure. The incompressible flow is assumed and we use the ideal gas law to express the relationship between density and temperature for natural convection [41]. As a result, the transport equations for incompressible turbulent flow can be written as follows:

Mass equation:

$$\frac{\partial \rho}{\partial t} + \frac{\partial}{\partial x_i}(\rho u_i) = 0 \tag{2.5}$$

Navier–Stokes equation:

$$\frac{\partial}{\partial t}(\rho u_i) + \frac{\partial}{\partial x_j}(\rho u_i u_j) = -\frac{\partial p}{\partial x_i} + \frac{\partial \tau_{ij}}{\partial x_j} + \rho g_i + F_i, \tag{2.6}$$

where the stress tensor τ_{ij} is defined as,

$$\tau_{ij} = \left[\mu\left(\frac{\partial u_i}{\partial x_j} + \frac{\partial u_j}{\partial x_i}\right)\right] - \frac{2}{3}\mu\frac{\partial u_i}{\partial x_i}\delta_{ij} \tag{2.7}$$

Energy equation:

$$\frac{\partial}{\partial t}(\rho E) + \frac{\partial}{\partial x_i}(u_i(\rho E + p)) = \frac{\partial}{\partial x_j}\left(k_{\text{eff}}\frac{\partial T}{\partial x_j} - \sum_j h_j J_j + u_j(\tau_{ij})_{\text{eff}}\right) + S_h, \tag{2.8}$$

where k_{eff} is the effective conductivity ($k + k_{\mathrm{t}}$, where k_{t} is the turbulent thermal conductivity); J_j is the diffusion flux of species j; S_{h} includes the heat of chemical reaction or any other volumetric heat sources. $(\tau_{ij})_{\mathrm{eff}}$ is the effective stress tensor, defined as:

$$(\tau_{ij})_{\mathrm{eff}} = \mu_{\mathrm{eff}}\left(\frac{\partial u_j}{\partial x_i} + \frac{\partial u_i}{\partial x_j}\right) - \frac{2}{3}\mu_{\mathrm{eff}}\frac{\partial u_{\mathrm{k}}}{\partial x_{\mathrm{k}}}\delta_{ij} \tag{2.9}$$

In Eq. (2.8):

$$E = h - \frac{p}{\rho} + \frac{v^2}{2}, \tag{2.10}$$

where sensible enthalpy h is defined for incompressible flows as:

$$h = \sum_j Y_j h_j + \frac{p}{\rho}, \tag{2.11}$$

where Y_j is the mass fraction of species j.

Equation for the turbulent kinetic energy k:

$$\frac{\partial}{\partial t}(\rho k) + \frac{\partial}{\partial x_i}(\rho k u_i) = \frac{\partial}{\partial x_j}\left[\left(\mu + \frac{\mu_{\mathrm{t}}}{\sigma_{\mathrm{k}}}\right)\frac{\partial k}{\partial x_j}\right] + G_{\mathrm{k}} + G_{\mathrm{b}} - \rho\varepsilon - Y_M + S_k \tag{2.12}$$

Equation for the energy dissipation:

$$\frac{\partial}{\partial t}(\rho\varepsilon) + \frac{\partial}{\partial x_i}(\rho\varepsilon u_i) = \frac{\partial}{\partial x_j}\left[\left(\mu + \frac{\mu_t}{\sigma_\varepsilon}\right)\frac{\partial\varepsilon}{\partial x_j}\right] + C_{1\varepsilon}\frac{\varepsilon}{k}(G_{\mathrm{k}} + C_{3\varepsilon}G_b) - C_{2\varepsilon}\rho\frac{\varepsilon^2}{k} + S_\varepsilon, \tag{2.13}$$

where G_{k} represents the generation of turbulence kinetic energy because of the mean velocity gradients and can be defined as $G_{\mathrm{k}} = -\rho\overline{u_i' u_j'}(\partial u_j/\partial x_i)$; G_{b} is the generation of turbulence kinetic energy due to buoyancy; σ_{k} and σ_ε are the turbulent Prandtl numbers for k and ε: $\sigma_{\mathrm{k}} = 1.0$, $\sigma_\varepsilon = 1.3$. $C_{1\varepsilon}$, $C_{2\varepsilon}$ are constants: $C_{1\varepsilon} = 1.44$, $C_{2\varepsilon} = 1.92$. $\mu_t = (C_\mu \rho k^2/\varepsilon)$ and $C_\mu = 0.09$.

The species (pollutants) transport equation:

$$\frac{\partial C_i}{\partial t} + u_j\frac{\partial C_i}{\partial x_j} = \frac{1}{\rho}\frac{\partial}{\partial x_j}\left[\left(D_i + \frac{\mu_{\mathrm{t}}}{Sc_{\mathrm{t}}}\right)\frac{\partial C_i}{\partial x_j}\right], \tag{2.14}$$

where C_i is the species (pollutants) concentration; D_i is the diffusivity; Sc_{t} is the turbulent Schmidt number.

The RNG approach, which is a mathematical technique that can be used to derive a turbulence model similar to the k–ε, results in a modified form of the epsilon equation, which attempts to account for the different scales of motion through changes to the production term. In addition, the RNG k–ε turbulence model has analytically derived differential formulas for determining the effective viscosity and the turbulent Prandtl number instead of a constant value for the standard k–ε turbulence model as follows:

$$\frac{\partial}{\partial t}(\rho k)+\frac{\partial}{\partial x_i}(\rho k u_i)=\frac{\partial}{\partial x_j}\left[\alpha_{\mathrm{k}}\mu_{\mathrm{eff}}\frac{\partial k}{\partial x_j}\right]+G_{\mathrm{k}}+G_{\mathrm{b}}-\rho\varepsilon-Y_M+S_{\mathrm{k}} \tag{2.15}$$

$$\begin{aligned}\frac{\partial}{\partial t}(\rho\varepsilon)+\frac{\partial}{\partial x_i}(\rho\varepsilon u_i)=&\frac{\partial}{\partial x_j}\left[\alpha_{\varepsilon}\mu_{\mathrm{eff}}\frac{\partial\varepsilon}{\partial x_j}\right]+C_{1\varepsilon}\frac{\varepsilon}{k}(G_{\mathrm{k}}+C_{3\varepsilon}G_{\mathrm{b}})\\&-\left[\rho C_{2\varepsilon}+\frac{C_{\mu}\rho\eta^3(1-\eta/\eta_0)}{1+\beta\eta^3}\right]\frac{\varepsilon^2}{k}+S_{\varepsilon},\end{aligned} \tag{2.16}$$

where α_{k} and α_{ε} are the inverse effective Prandtl number for k and ε, respectively; μ_{eff} is the effective turbulent viscosity; $\eta=Sk/\varepsilon$, S is the scalar measure of the deformation tensor; the constants of η_0 and β are 4.38 and 0.012, respectively.

For two-dimensional model, the temperature at land surface is assumed constant and higher than that of incoming wind flow, which could be valid due to strong solar radiation and low heat capacity of urban road materials. For three-dimensional model, solar radiation will be imposed on the ground surface and all the building surfaces, varying with time. As shown in Fig. 2.2, solar radiation (I_c) on an arbitrary tilted surface having a tilt angle of β from the horizontal and an azimuth angle of α_w is the sum of components consisting of beam ($I_{b,c}$), sky-diffused ($I_{d,c}$) and ground-reflected solar radiation ($I_{r,c}$) [42, 43]:

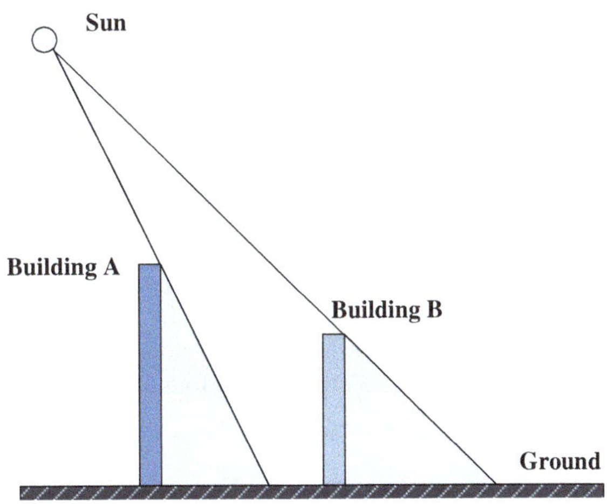

Fig. 2.2 Schematic of solar radiation on buildings

$$I_c = I_{b,c} + I_{d,c} + I_{r,c} \tag{2.17}$$

$$I_{b,c} = I_{b,N} \cos i = I_{b,N}[\cos\alpha \cos(\alpha_s - \alpha_w) \sin\beta + \sin\alpha \cos\beta] \tag{2.18}$$

$$I_{d,c} = CI_{b,N} \cos^2\left(\frac{\beta}{2}\right) \tag{2.19}$$

$$I_{r,c} = \gamma I_{b,N}(\sin\alpha + C) \sin^2\left(\frac{\beta}{2}\right) \tag{2.20}$$

$$I_{b,N} = C_n I \mathrm{e}^{-L/\sin\alpha}, \tag{2.21}$$

where C, C_n are constants related to the sky conditions, L is the optical depth for average atmospheric condition at sea level with a moderately dusty atmosphere, and γ is the surface reflectance.

In Eq. (2.21), the solar radiation I can be calculated as:

$$I = I_0\left(\frac{D_0}{D}\right)^2, \tag{2.22}$$

where I_0 is the solar constant, D is the distance between the sun and the earth, and D_0 is the yearly mean earth–sun distance. The factor $(D_0/D)^2$ can be approximated as follows with very high accuracy [43]:

$$\left(\frac{D_0}{D}\right)^2 = 1.00011 + 0.034221 \cos x + 0.00128 \sin x + 0.000719 \cos(2x) + 0.000077 \sin(2x) \tag{2.23}$$

$$x = \left(\frac{360(n-1)}{365}\right)^0, \tag{2.24}$$

where n is day number starting from January 1 as 1.

Another key issue for the solar radiation is the calculation of shadow angle. For example, there are two buildings within a short distance of each other (Fig. 2.2). If the sun lies in position A, two shadows will occur, with the right building also being influenced by the left building. All the surfaces under the shadows will receive no direct solar radiation at this time but still can receive sky-diffused radiation and ground-reflected radiation. On the other hand, if the sun lies in position B, all the left building is under the shadow of the right building. In this way, the shadow angle and view factors among the surfaces should be seriously considered, with calculation theory found in Ref. [42], and view factors (sometimes called shape factor) calculation theory easily found in any heat transfer textbook.

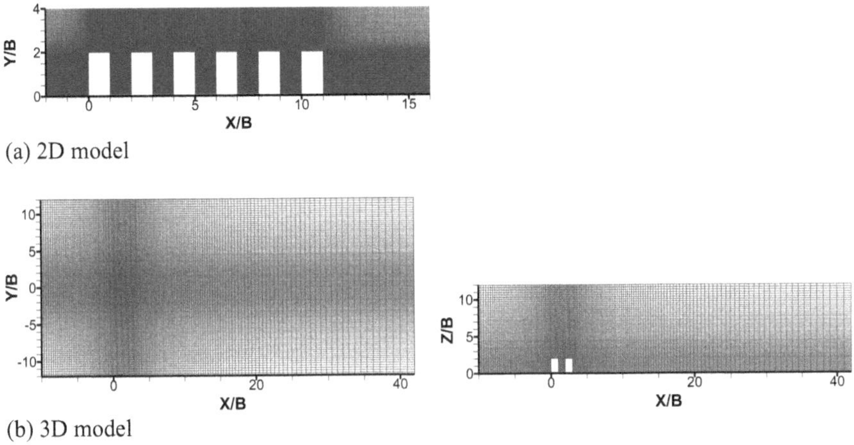

Fig. 2.3 Grid distribution of the geometric models

2.4 Meshing Skills

In general, for the same meshing zone, the hexahedral (HEX) meshing method is more economical and can reduce false diffusion more efficiently than the tetrahedral method. As a result, HEX grids were applied in the model and the mesh generation procedure of the whole geometric model was executed using the commercial software package Gambit 2.4.6. Figure 2.3 reveals the grid distribution of the geometric models. Because of the anticipated relatively steep gradients in velocity, pressure, and temperature near the walls and the ground, serious consideration should be given to the boundary layer effect of these areas. Thereby, the fine grids near the building walls and ground surfaces were more concentrated in these locations.

2.5 Computational Method

The computations have been performed by using the general-purpose CFD program ANSYS Fluent 14.0. The QUICK scheme was used to discretize the convective terms and a second order accurate treatment was used for the diffusion terms. The discretized algebraic equations were solved in a coupled manner. The iterations were continued until the relative error in the mass conservation equation was below 1×10^{-6} and in the energy equation was below 1×10^{-8}.

2.6 Verification

The grid-independent performance of the numerical simulation results were verified between some test cases with different grid numbers, the results show that there was only a deviation of approximately 1.0% between these compared results, which demonstrated the solutions in this article are grid-independent. To test the validity of the mathematical model describing the fluid flow, heat transfer characteristics of a certain system, the most feasible method is to compare the numerical results with the wind tunnel experimental results [44] by being given the same boundary conditions and working conditions. Figure 2.4 denotes the numerical velocity and temperature profiles with experimental results at street canyon centerline. The streamwise velocity magnitude ($U_{\times}$) is divided by the wind velocity at building height (U_{ref}) to calculate normalized streamwise velocity. Meanwhile, the normalized temperature is defined by the formula that the difference between air and ground temperature ($T - T_0$) is divided by the surface and ambient air temperature difference ($T_{\mathrm{ref}} - T_0$). It is obvious that, the normalized potential temperature agrees well with the wind tunnel data in the street canyon, but the normalized horizontal velocity has not followed that better and the deviations above the street canyon are slightly large than those within the street canyon. The temperature profiles generally agree well with the data obtained from the wind tunnel experiment. Despite some slight differences, the numerical codes and procedures applied in the chapter is feasible.

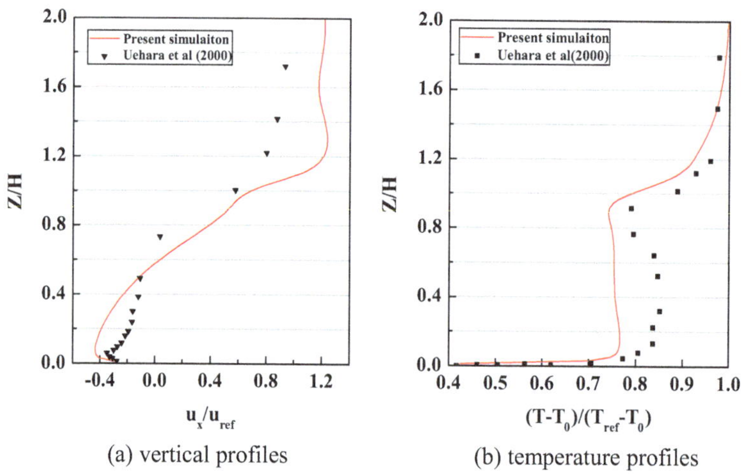

Fig. 2.4 Numerical velocity and temperature profiles with experimental results at street canyon *centerline*

2.7 Results and Discussion

2.7.1 Comparison of 2D Flow Characteristics

Figures 2.5, 2.6, 2.7 and 2.8 displays the 2D flow characteristics at the width-to-height ratio, $H/B = 2$ when the source strength of pollutant is 5 kg/(m^3 s). When the inflow wind approaches the first canyon in the windward side, a strong elevated trailing vortex system is generated, and there exists flow separation and strong turbulence near the top windward corner of the first canyon. The results obtained indicate that the mean horizontal velocity was very small within the street canyon due to the sheltering effect of buildings, thus a strong shear layer presented below the canyon canopy, as shown in Fig. 2.5. Airflow within the canyons is supposed to be slow and the ventilation may be depended upon the vortex circulation. Thus the vertical velocity within the canyons induced by the vortex plays an important role on the air exchange between the street canyons and the external flow above street canopy. In addition, turbulence help turbulent mixing within the street canyons also contribute to air exchange and pollutant removal, but the lowest turbulent kinetic energy was found at the separation and reattachment points associated with the corner of the down part of the street canyons. In contrast, the turbulent kinetic energy is relatively large near the canyon roof level and vertical fluxes thus induced, and it increases with the increasing wind speed. As shown in Fig. 2.8, the vortex circulation within the street canyon shows a similar behavior, two contrarotative vortices are generated within the each street canyon when the approaching wind is blown across the canyons. It shows that this behavior belongs to the multi-vortex regime as described by Refs. [8, 9] in such deep street canyons ($H/B > 1.67$). Besides, a strong vortex and a tiny vortex are generated at the corner region on the leeward of the street canyons. As a result, recirculation flows induced

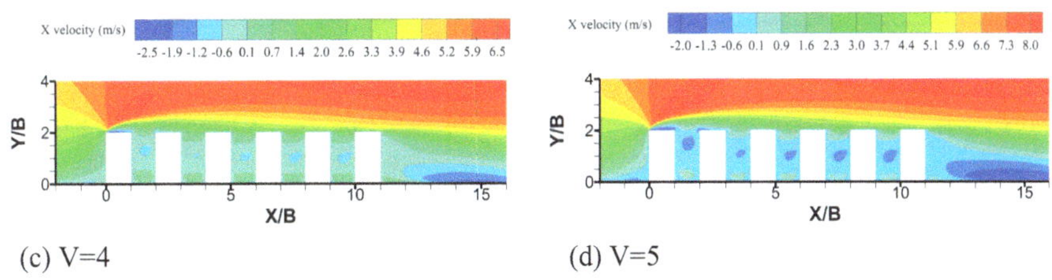

Fig. 2.5 Stream-wise (*X* direction) velocity distributions under different approaching wind speeds

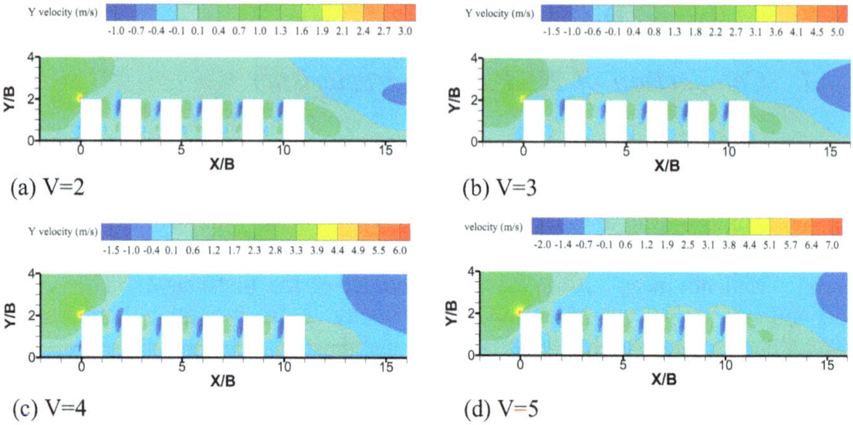

Fig. 2.6 Vertical (*Y* direction) velocity distributions under different approaching wind speeds

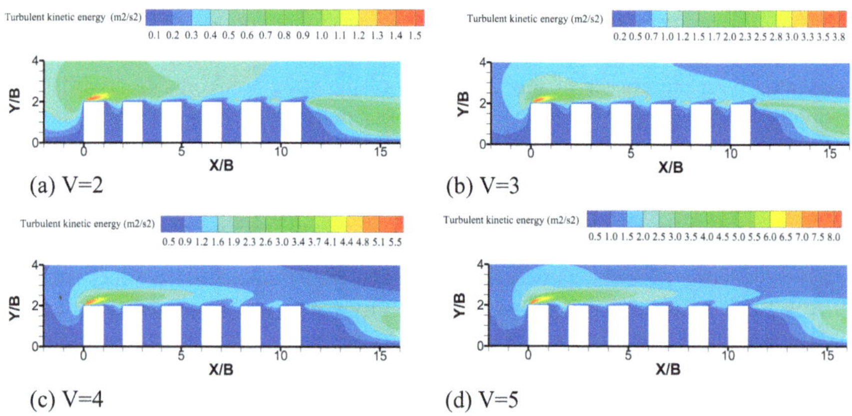

Fig. 2.7 Turbulent kinetic energy distributions under different approaching wind speeds

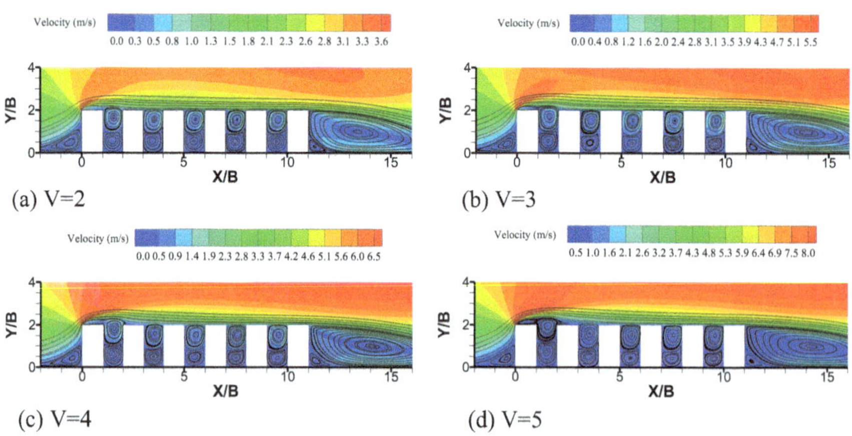

Fig. 2.8 Velocity distributions and streamlines under different approaching wind speeds

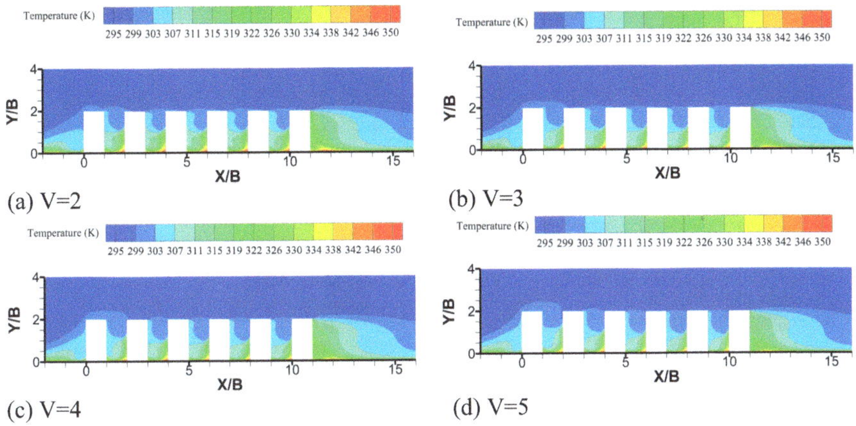

Fig. 2.9 Temperature distributions under different approaching wind speeds

by those vortexes result in poor capacity of dispersing ground-level pollutants, thus the traffic-related pollutants are difficult to be removed out through these deep canyons.

Figure 2.9 shows the temperature distributions under different approaching wind speeds at the width-to-height ratio, $H/B = 2$. It is clear that bottom heating could induce the thermal stratification, which is similar to the shear layer exists in the near field of urban canopy layer. Moreover, as a multi-vortex SF regime formed, two contrarotative vortex circulations are able to have an impact on the temperature distributions within the street canyon. The upper vortex has a larger rotative speed, whereas the lower vortex circulation has the smaller rotative speed, the capacity of air exchange within the street canyon is poor. Hence, the results show that the street canyon with high-rise buildings may suppress the air ventilation, leads to a direct impact on human health. Meanwhile, it also found that temperature distribution is independent on approaching wind speed shown in the Fig. 2.8, under the circumstance that temperature at ground surface is assumed constant.

Figure 2.10 demonstrates the mass fraction of pollutant (CO) under different approaching wind speeds within the street canyon for $H/B = 2$ and the source strength of pollutant is 5 kg/(m^3 s). As the circulation of two contrarotative vortices generated within the canyon, the pollutant dispersion mechanism within the canyon is fully dependent on the recirculation flows. The upper clockwise with a larger rotative speed could transport the pollutant to the leeward side of the buildings, thus the pollutants accumulates at the leeward wall of the buildings. Besides, the airflow at the bottom of the upper vortex blocked by the wall surface then diverges downwardly. Thus the counterclockwise vortex circulation with the smalled rotative speed formed at the lower region of the street canyon. It can carry the pollutant released from the ground pollutant source zone to the windward side of downstream building. Accordingly, we can see more pollutants accumulate on the windward side at the lower region of the street canyon. And because of the small rotative

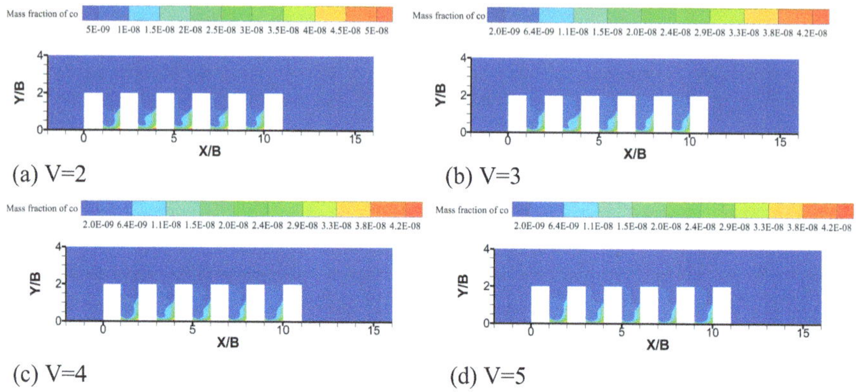

Fig. 2.10 Mean concentrations of pollutant under different approaching wind speeds

speed of lower vortex circulation, the capacity of pollutant removal in these deep street canyons is very poor. From the result we can see that the simulated pollutant concentrations are less dependent on the variation of wind speed for the studied street canyon aspect ratio of $H/B = 2$ case.

To investigate the effect of source strength of pollutants on dispersion mechanism within the street canyon, Fig. 2.11 shows the normalized concentrations of pollutant under different source strength of pollutant, $Q = 5, 10, 15, 20$ kg/(m^3 s), and the reference approaching wind speed is 2 m/s. It can be seen that distributions of the normalized dimensionless pollutant concentrations of different source intensities are almost the same as the mean concentrations of different reference wind speeds, whether on the leeward side of the building and windward side of the building. As we discussed above, it is because that the recirculation flows induced within the street canyons. In addition, as K is inversely proportional to the source strength, Q. Indeed, the mass fractions of pollutant for these cases increase with increasing source strength Q. However, it is safe to draw a conclusion that source strength has little impact on the normalized dimensionless pollutant concentrations

Figure 2.12 demonstrates the streamwise (X direction) velocity distributions under different street widths (i.e., $H/B = 2$, $H/B = 1.43$, $H/B = 1.11$, $H/B = 0.91$, $H/B = 0.77$) when the source strength of pollutant is 5 kg/(m^3 s) and the reference wind speed is 2 m/s. As we can see from Fig. 2.12a–e, there is a flow separation near the top windward corner of the first canyon, a strong shear layer always exits in the canopy layer. It is confirmed that the flow field in the urban canopy layer is perturbed by building obstacles, and is characterized by the interaction between atmospheric flow and flow over street canyons. As we observe the airflow patterns within the street canyons, it is apparently that there are recirculation flows of vortices inside the canyons, and the position of these vortices vary with the street widths. The wider the street width, the lower region the vortices existed. As mentioned before, the air ventilation or pollutant dispersion within the street canyons depend largely on the vortex circulations. Thus the features of vortex

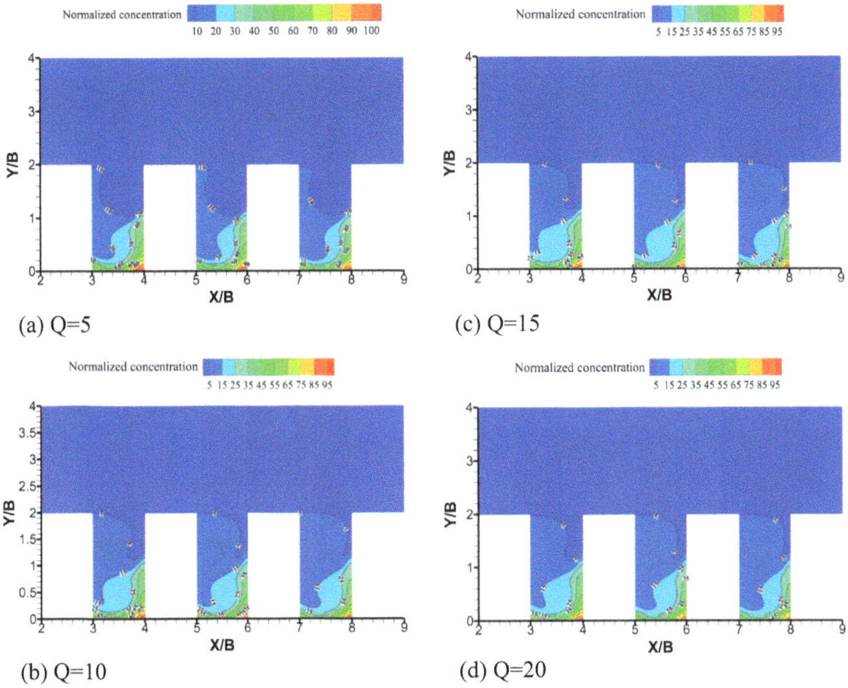

Fig. 2.11 Normalized concentration of pollutant under different source strength of pollutants

circulations shown in the figure represent the flow pattern of airflow within the street canyons, have different impacts on the air ventilation or pollutant dispersion. Indeed, when the position of vortex circulation is lower, the entrainment of the pollutant which released from the ground source zone would be easier and is conductive to air exchange with the external airflow. As a result, with the increase of the street width, the air ventilation within the canyon is better.

Figure 2.13 shows the vertical velocity distributions under different street widths when the source strength of pollutant is 5 kg/(m^3 s) and the reference wind speed is 2 m/s. It is clear that the maximum value of vertical velocity presented at the top windward corner of the first building, due to the approaching wind blocked by the wall surface of the first building. Meanwhile, we can see there is a vortex appears on the windward wall surface from the second building to the sixth building. As the vertical mixing contributes to the air exchange between the air within the canyons and the external airflow, the size and strength of these vortexes largely determine the air ventilation within the street canyons.

Figure 2.14 shows the turbulent kinetic energy distributions under different street widths when the source strength of pollutant is 5 kg/(m^3 s) and the reference wind speed is 2 m/s. Similar to the vertical velocity, the maximum value of the turbulent kinetic energy appears near the top windward corner of the first building due to the inflow wind approaches the wall surface of the first building. And we can

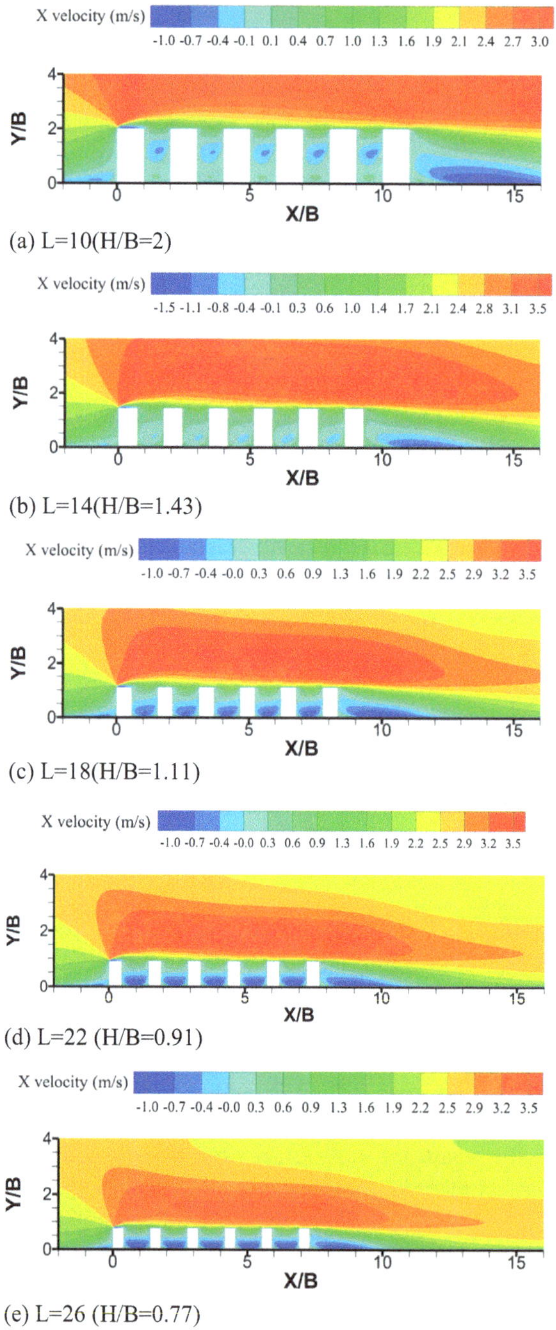

Fig. 2.12 Streamwise (*X* direction) velocity distributions under different street widths

Fig. 2.13 Vertical (Y direction) velocity distributions under different street widths

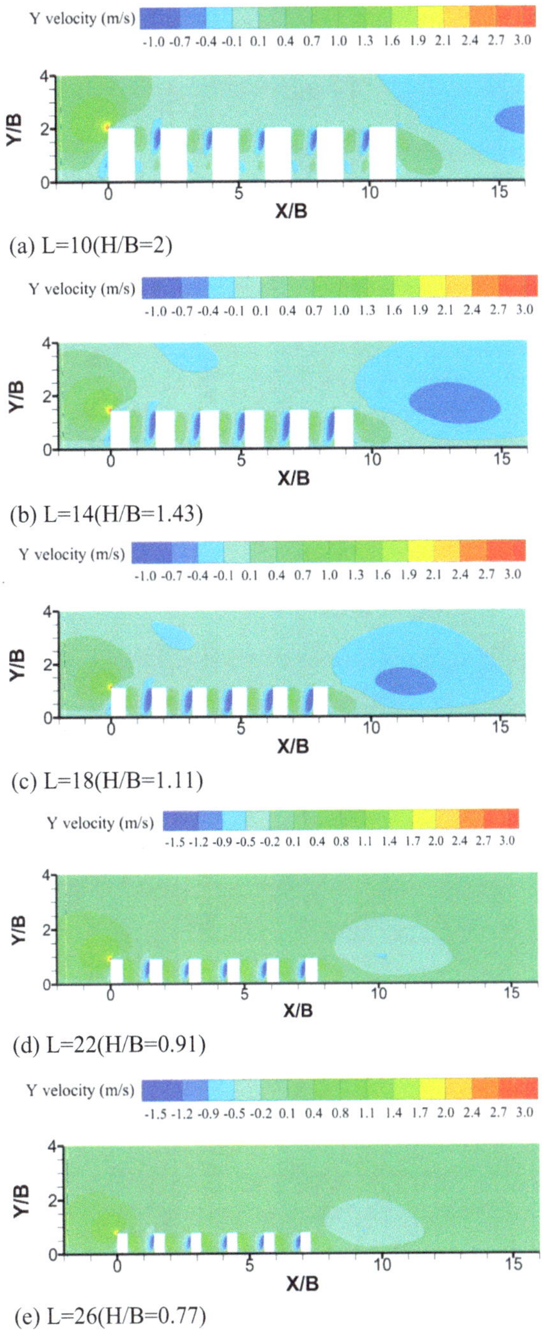

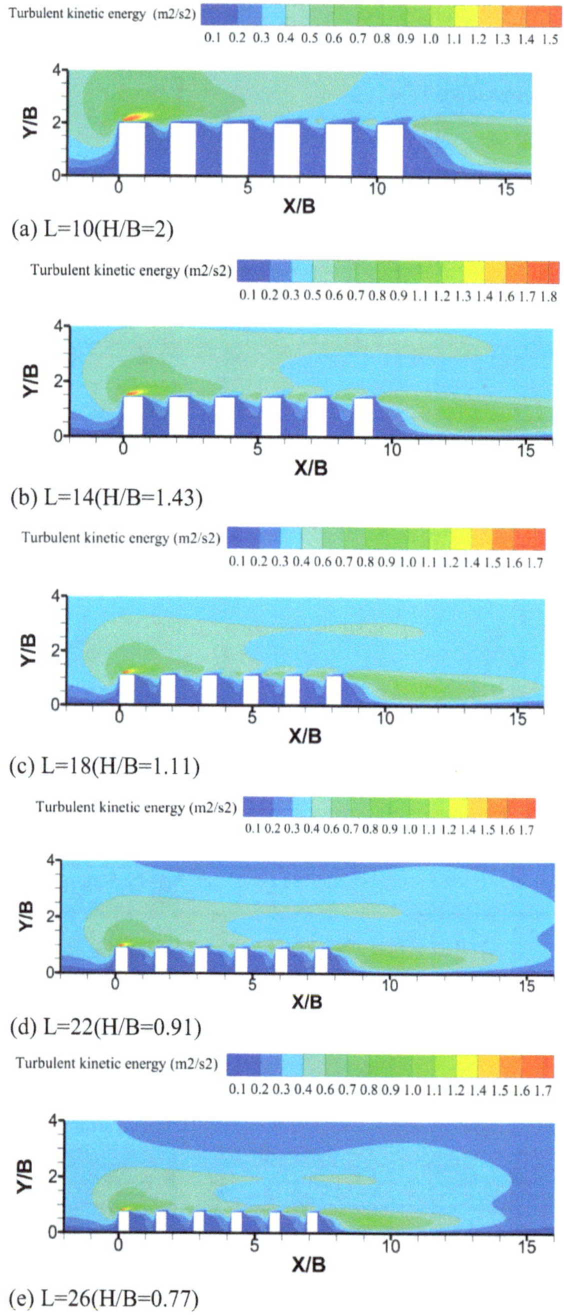

(a) L=10(H/B=2)

(b) L=14(H/B=1.43)

(c) L=18(H/B=1.11)

(d) L=22(H/B=0.91)

(e) L=26(H/B=0.77)

Fig. 2.14 Turbulent kinetic energy distributions under different street widths find that there is almost no significant change on the distribution of turbulent kinetic energy for different street widths or street aspect ratios, but with the increase of street widths, the turbulent kinetic energy within the street canyons is bigger. This means that the turbulent mixing within the street canyons is stronger resulted in better air exchange and ventilation with the wider street width.

The street aspect ratio has significant influence on the flow field and pollutant dispersion from the traffic emission source. Figure 2.15 demonstrates the velocity distributions and streamlines under different street widths or street aspect ratios, i.e. $H/B = 2$, $H/B = 1.43$, $H/B = 1.11$, $H/B = 0.91$, $H/B = 0.77$, when the source strength of pollutant is 5 kg/(m^3 s) and the reference wind speed is 2 m/s. It is clearly shown that flow patterns of approaching wind are quite different by comparing the studied street aspect ratios from $H/B = 2$ to $H/B = 0.77$. For the $H/B = 2$ case, two contrarotative vortices are generated within the each street canyon when the approaching wind is blown across the canyons which belongs to the multi-vortex regime. A strong elevated trailing vortex system is generated at the corner region on the windward of the street canyons, whereas a strong vortex and a tiny triangular vortex are generated at the corner region on the leeward of the street canyons. With the increase of the width of the street, there is no significant variation of flow patterns between these studied cases. However, the biggest change that can be seen is that the flows or vortexes distribution within the street canyons. It seems that the flow regime within the street canyons varies from multi-vortex regime to SF with the increase of street width. As we can see Fig. 2.15b–e, there is a strong vortex exists within the street canyons, which contributes to the air ventilation and pollutant removal. Despite there are also some tiny vortexes at the corner of windward wall surface and at the corner of leeward wall surface, being not really obvious with the increasing street widths, it is a typical SF.

Figure 2.16 displays the temperature distributions under different street widths or street aspect ratios, i.e. $H/B = 2$, $H/B = 1.43$, $H/B = 1.11$, $H/B = 0.91$, $H/B = 0.77$, when the source strength of pollutant is 5 kg/(m^3 s) and the reference wind speed is 2 m/s. From Fig. 2.15a–e, bottom heating could induce the thermal stratification, but there are also some differences between these studied cases. On one hand, the temperature at the corner of windward surface wall of buildings seems larger when $H/B = 2$. It is mainly because the two contrarotative vortices of the multi-vortex regime, lower vortex circulation with low speed transports more air or pollutants to the windward surface wall of building, thus here is the worst ventilation. On the other hand, with the increase of street width, the temperature distribution becomes more uniform, the temperature at corners of the windward surface wall and of the leeward surface wall of building is not too high, and this means that better air ventilation within the street canyons. In conclusion, with the increase of street canyon, the temperature distribution within the street canyons becomes more uniform, air ventilation is getting better.

Figure 2.17 shows the normalized concentration of pollutant under different street widths or street aspect ratios, i.e. $H/B = 2$, $H/B = 1.43$, $H/B = 1.11$, $H/B = 0.91$, $H/B = 0.77$, when the source strength of pollutant is 5 kg/(m^3 s) and the reference wind speed is 2 m/s. It is obvious that, with the increase of the width

Fig. 2.15 Velocity distributions and streamlines under different street widths

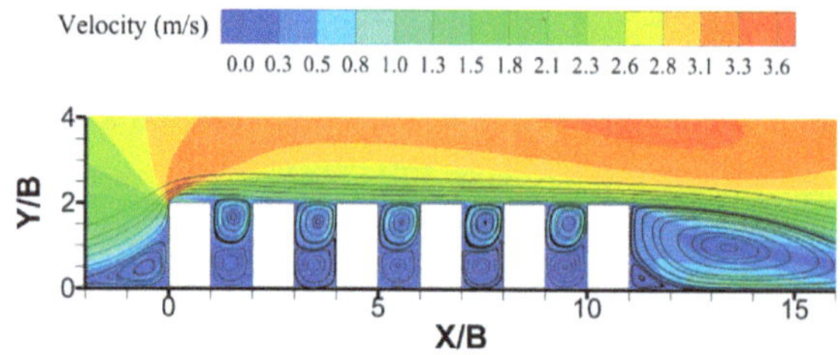

(a) L=10(H/B=2)

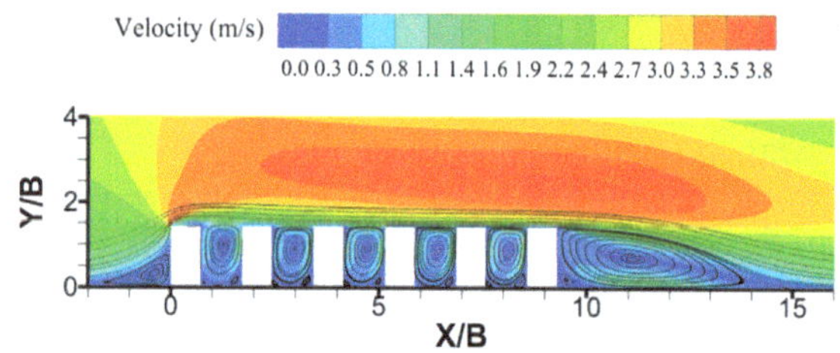

(b) L=14 (H/B=1.43)

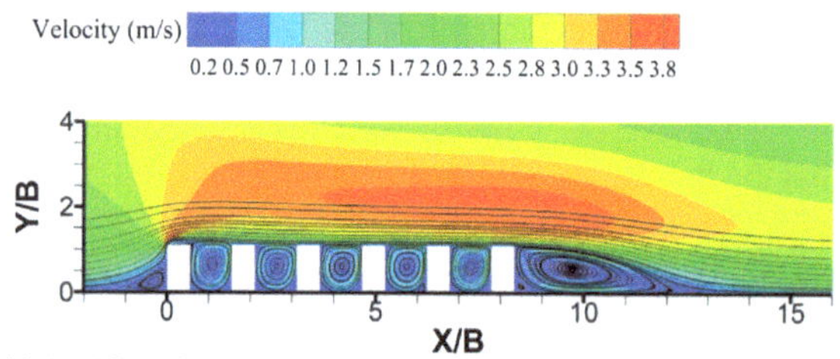

(c) L=18 (H/B=1.11)

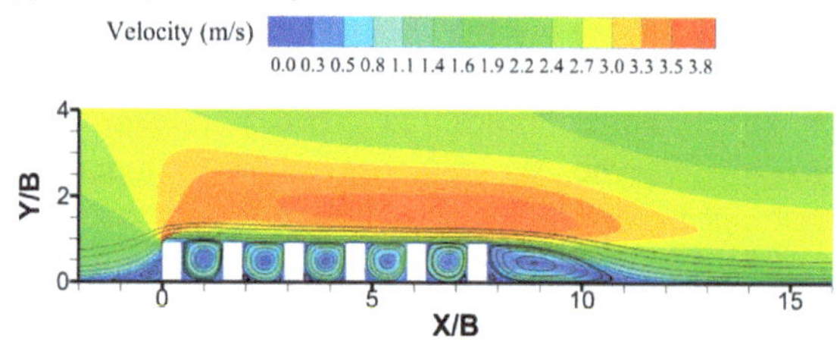

(d) L=22 (H/B=0.91)

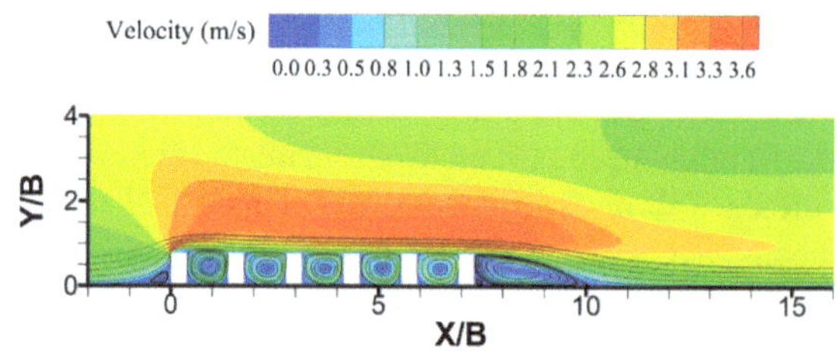

(e) L=26 (H/B=0.77)

Fig. 2.16 Temperature distributions under different street widths

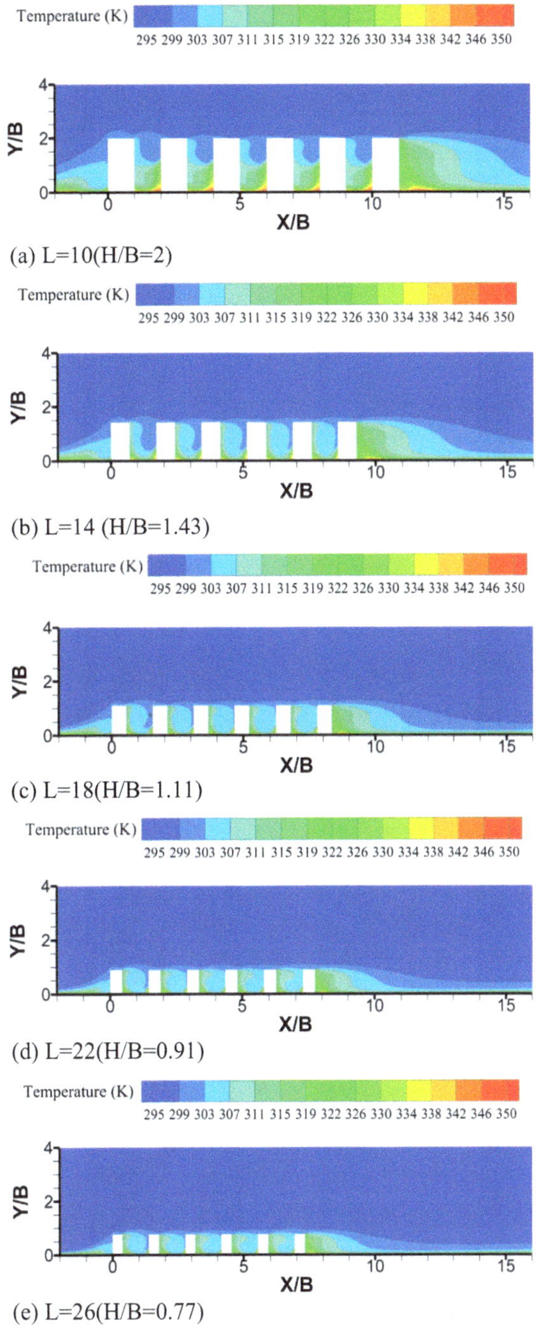

(a) L=10(H/B=2)

(b) L=14 (H/B=1.43)

(c) L=18(H/B=1.11)

(d) L=22(H/B=0.91)

(e) L=26(H/B=0.77)

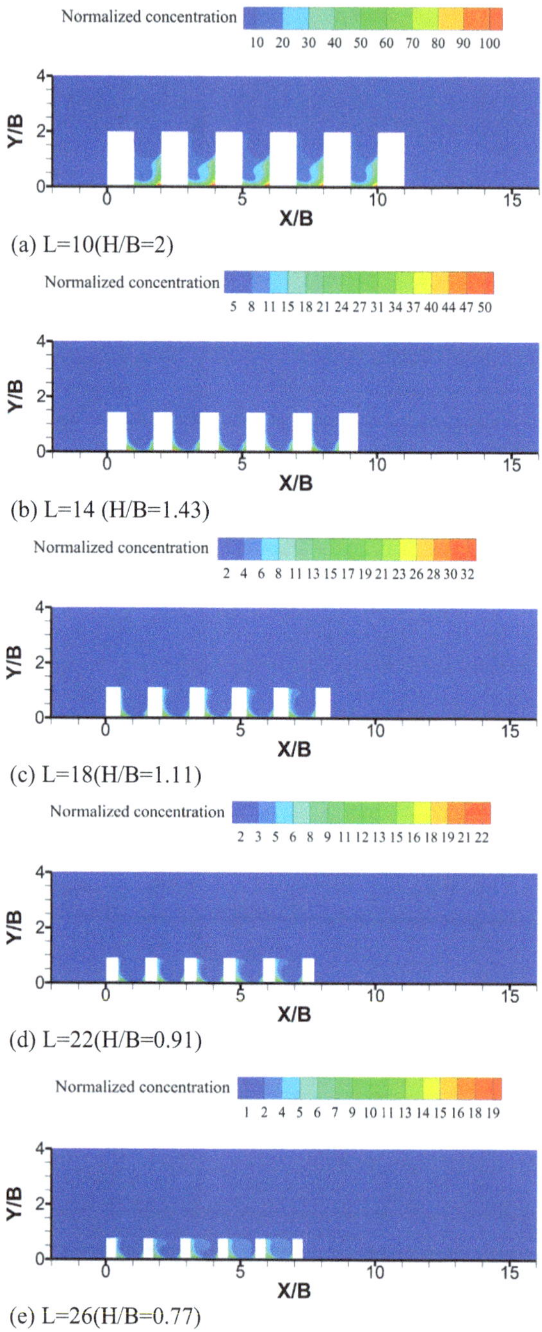

Fig. 2.17 Normalized concentration of pollutant under different street widths

of the street, pollutant concentrations continue to decline within the street canyons. The only difference is the position of pollutant accumulated from these studied cases. As shown in Fig. 2.17a, more pollutant accumulated near the windward side of street canyon due to the vortices of the multi-vortex regime in such deep street canyons. Conversely, pollutant would like to accumulate near the leeward side of street canyon more than the windward side with the increasing street widths. In fact, this phenomenon is caused by vortex circulations within the street canyons. With the increase of street width, the vortex flow patterns within the street canyons are changed accordingly. As a result, wider street can promote air ventilation within the street canyon, hence, less pollutant is accumulated.

To investigate the flow characteristics in street canyon with a building height variation, Fig. 2.18 displays the streamwise (X direction) velocity distributions between two studied cases when the source strength of pollutant is 5 kg/(m^3 s) and the reference wind speed is 2 m/s. As shown in figure, the approaching wind seems to be blocked by the windward top corner of the first taller building, and then the stagnated airflows diverge into two different directions, flowing upwardly and flowing downwardly. When taller buildings located behind the shorter buildings, it seems that the magnitude of streamwise velocity at the shorter building level is greater than that of the taller buildings located in front of the shorter buildings. At the same time, we find that various strong vortices are easier to be formed behind the taller buildings from these two studied cases, which contribute to turbulent mixing but not mean air ventilation at this place is better. Instead, it may worsen the air exchange between the street canyons with external airflow by increasing the

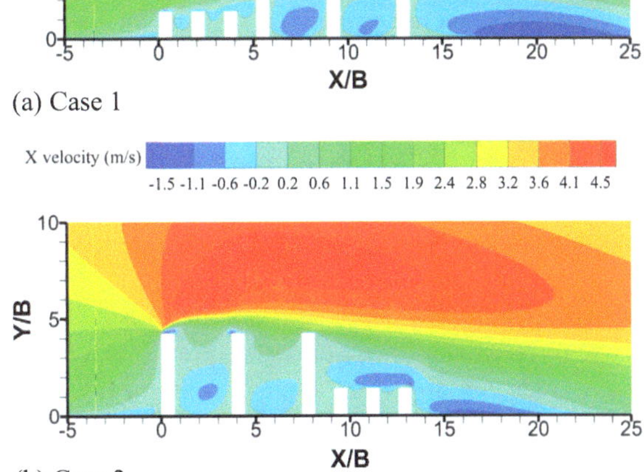

(a) Case 1

(b) Case 2

Fig. 2.18 Stream-wise (X direction) velocity distributions between two studied cases

Fig. 2.19 Vertical (*Y* direction) velocity distributions between two studied cases

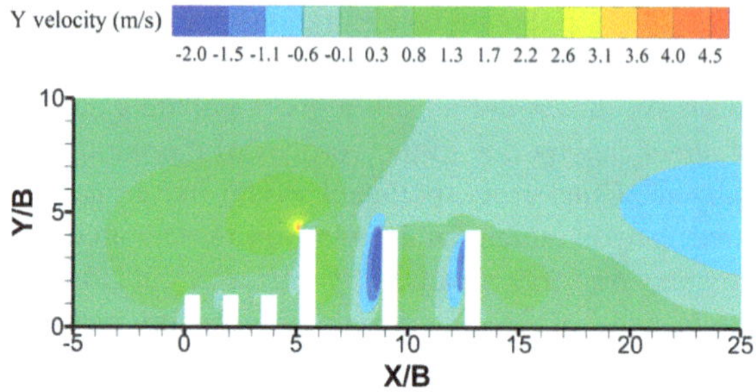

(a) Case 1

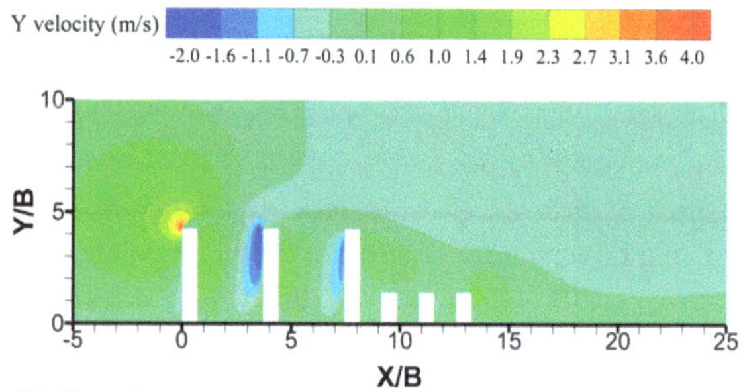

(b) Case 2

level of canopy layer. Thus it is tempting to conclude that air ventilation in front of the taller buildings may be better than those behind the taller buildings.

Figure 2.19 displays the vertical (*Y* direction) velocity distributions between two studied cases when the source strength of pollutant is 5 kg/(m^3 s) and the reference wind speed is 2 m/s. The maximum value of the vertical velocity both appears at the windward top corner of the first taller building, it means that the blocking effect of buildings to approaching wind flow is the largest at this place. In street canyons behind the taller buildings, the vertical velocity is mainly positive, which means the wind mainly flows upwardly and flows over the buildings. On the other hand, the large negative velocity appears in front of the latter two taller buildings, which means the wind flows downwardly and vortex circulations are often formed at these places. Hence it may be difficult to distinguish which is the better case for air ventilation just from Fig. 2.19.

Figure 2.20 shows the turbulent kinetic energy distributions between two studied cases when the source strength of pollutant is 5 kg/(m^3 s) and the reference wind speed is 2 m/s. The maximum value of the turbulent kinetic energy both appears at the windward top corner of the first taller building in two cases, which means that the turbulent mixing is the largest at this place. Besides, it shows that the influenced scope of turbulent kinetic energy is greater when taller buildings located in front of the shorter buildings, to some extent means that the drag force produced by buildings is large when wind is blown from windward to leeward direction, which may weaken air ventilation within the street canyons.

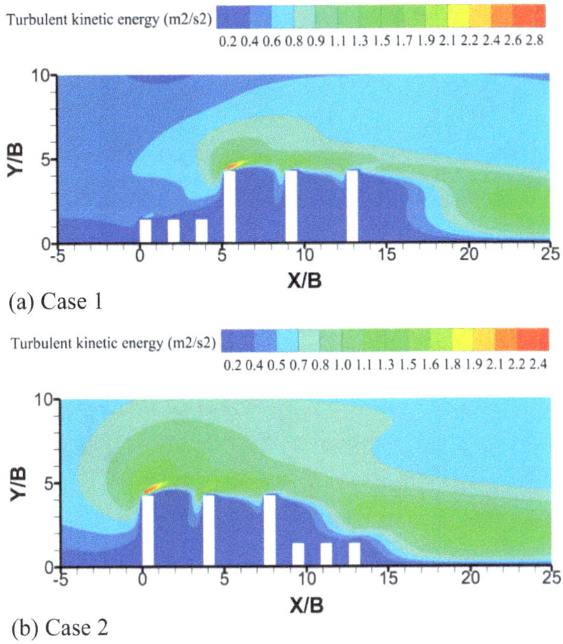

Fig. 2.20 Turbulent kinetic energy distributions between two studied cases

Figure 2.21 demonstrates the velocity distributions and streamlines between two studied cases when the source strength of pollutant is 5 kg/(m^3 s) and the reference wind speed is 2 m/s. It can be found that the flow patterns within street canyons characterized by various vortex circulations. In general, the distributions of vortex flows within the street canyons of taller buildings are almost same from two studied cases. The difference is that the vortices distributions in front of the taller buildings and behind the taller buildings which may determine the air ventilation within street canyons between these two studied cases. In case 1, where the taller buildings located behind the shorter buildings, the level of canopy layer of these shorter buildings is lower compared to that of case 2. While it does not mean case 1 layout has a direct positive effect on the air ventilation within the street canyons, but makes it easier for air exchange between the air inside the canyons and the external flows. In addition, we also notice that the shadow behind the taller buildings is always large, filled with vortex circulation, which may result in poor ventilation here. Therefore, compared to the two kinds of buildings layout, arranging the taller buildings behind the shorter buildings may obtain relatively better air ventilation.

Figure 2.22 displays the temperature distributions between two studied cases when the source strength of pollutant is 5 kg/(m^3 s) and the reference wind speed is 2 m/s. As can be seen from the figure, the temperature distributions within the taller buildings are almost same between these two studied cases. However, the temperature distributions in front of the taller buildings and behind the taller buildings are clearly different. That is, the local high temperature would like to appear behind the taller buildings, which is due to the blocking effect by taller buildings is greater

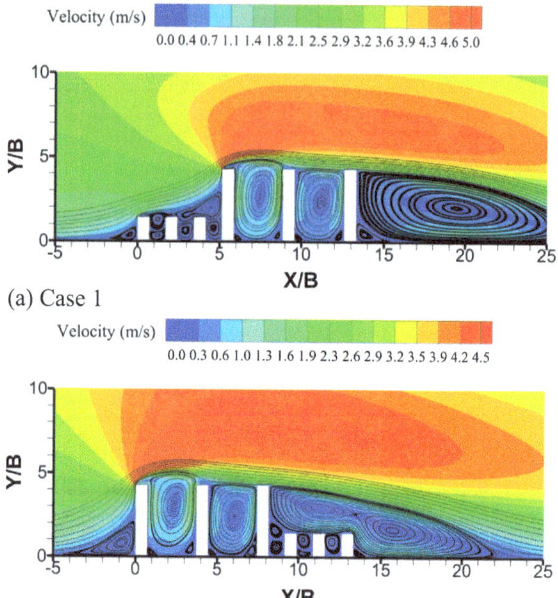

(a) Case 1

(b) Case 2

Fig. 2.21 Velocity distributions and streamlines between two studied cases

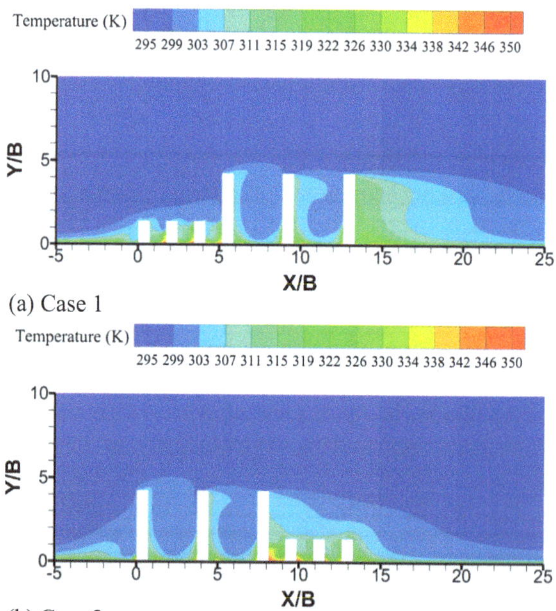

(a) Case 1

(b) Case 2

Fig. 2.22 Temperature distributions between two studied cases

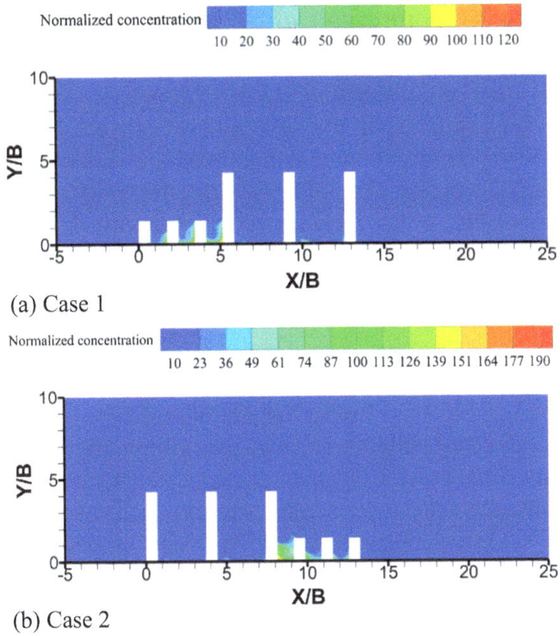

Fig. 2.23 Normalized concentration of pollutant between two studied cases

than that of the shorter buildings. Thus air ventilation at this place seems weak, heat or pollutants are difficult to be removed. Moreover, in case 2, it shows that the local temperature within the street canyon is much higher than that of case 1 behind the taller buildings, which implies high degree of heat or pollutants are accumulated here then pose a direct threat to human health.

Figure 2.23 shows the normalized concentration of pollutant between two studied cases when the source strength of pollutant is 5 kg/(m^3 s) and the reference wind speed is 2 m/s. Because of the wider street width between the taller buildings, the level of normalized concentration is relatively low within the canyons. In contrast, the concentrations are great between the shorter buildings thus pollutants are more easily accumulated here. On one hand, it is due to that the narrow street causes more vortex circulations with low rotation speed which not good to air ventilation. On the other hand, it is mainly because the great blocking effect by the taller buildings, thus a small amount of air can enter the street canyons between the shorter buildings. Overall, the pollutants distributions corresponding to the flow field as we analyzed before. In case 1, when the taller buildings located behind the shorter buildings in the direction of approaching wind, the air exchange between the street canyons with external airflows is better than that of the case 2, concentrations of contamination and heat are easier to be removed. However, concentrations of pollutants and heat within the street canyons behind the taller buildings are difficult to mitigate.

2.7.2 Comparison of 3D Flow Characteristics

As we mentioned before, pollutant dispersion is controlled by turbulent mixing and mean flow transport through the street canyon. The realistic flow structure around a building is highly three-dimensional and complex when the wind approaches the building can be identified by considering bluff body aerodynamics and atmospheric dispersion, which includes curved streamlines, velocity discontinuities, and non-homogeneous turbulence disperse effluents. Hence it is necessary to reproduce three-dimensional flow to accurately predict the flow and pollutant dispersion around street canyons.

Figure 2.24 demonstrates the 3D streamlines under different approaching wind speeds within the street canyon for $H/B = 2$ and the source strength of pollutant is 5 kg/(m^3 s). It is obvious that, the approaching wind is blocked by the windward top corner of the building, and then the airflows diverge into three directions, i.e., flowing upwardly, flowing downwardly and flow laterally. The separated and reattached roof flows are found over the street canyons, some air flows downwardly into the canyon, vortex circulation within the canyon thus formed. Besides, it shows that vortexes are also can be found behind the buildings, despite the magnitude of rotation speed is relatively low which implies the capacity of air ventilation here is a little poor. Of course, when the approaching wind speed increases, the ventilation conditions behind the buildings are improved to a certain extent, but not really evident. In addition, the flow patterns of mean flows are almost same with the increasing wind speed, which indicates that the 3D flow characteristics are less dependent upon the approaching wind speed. However, the turbulence characteristics around two buildings and street canyons are not fully reproduced in RANS approaches, such as the horseshoe vortex of the lateral flows and vortex shedding reported by Ref. [5].

The temperature at land surface in two-dimensional model is assumed constant and higher than that of approaching airflow, but the more realistic situation is that due to strong solar radiation and heat capacity of urban road or building materials, solar radiation will be imposed on the ground surface and all the building surfaces, varying with time. Therefore we choose a particular solar time given the same solar radiation conditions to study the temperature distributions under different approaching wind speeds within the street canyon for $H/B = 2$ and the source strength of pollutant is 5 kg/(m^3 s), as shown in Fig. 2.25. Overall, there is no apparent change on the temperature distributions of ground surface and building surfaces. The only difference is that, the mean temperature is relatively high when the approaching wind speed is 2 m/s, which leads to poor natural ventilation compared to other cases. As a result, the place in the windward direction of buildings with a low mean temperature at ground level, whereas the place in the leeward direction of buildings with a comparative high mean temperature at ground level. In addition, the temperature distribution within the street canyon, the closer the middle of street and the closer low region of canyon, the local high temperature is likely existed.

Due to the flow field and the temperature field, the pollutant dispersion within the street canyon has a fully three-dimensional characteristic. Thus Fig. 2.26

Fig. 2.24 3D streamlines under different approaching wind speeds

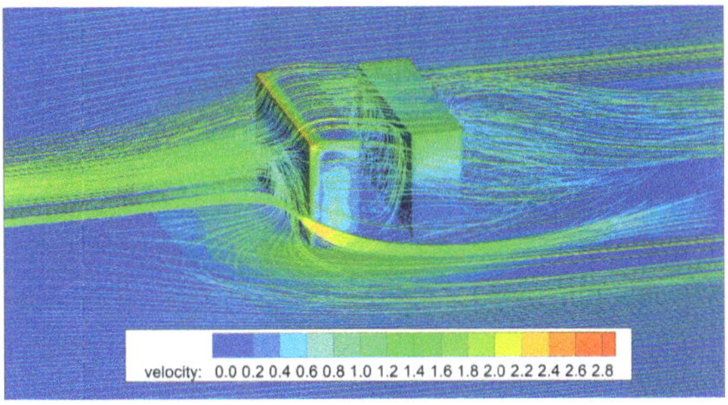

(a) V=2

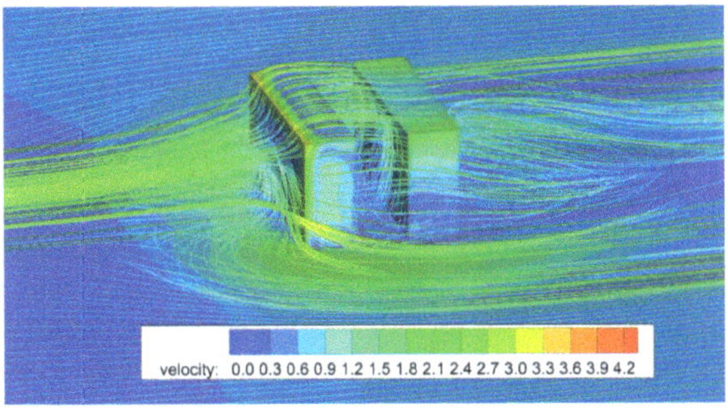

(b) V=3

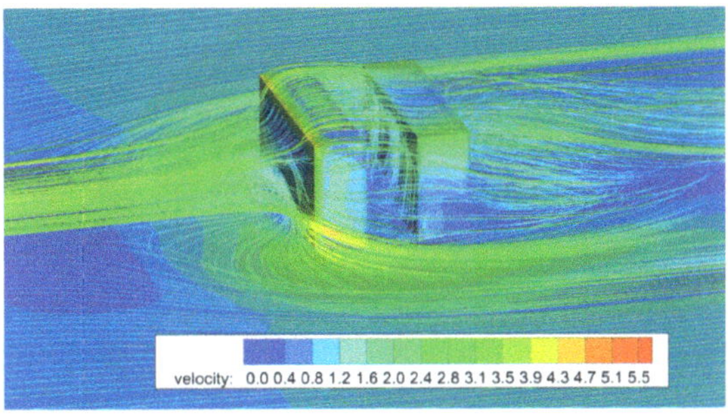

(c) V=4

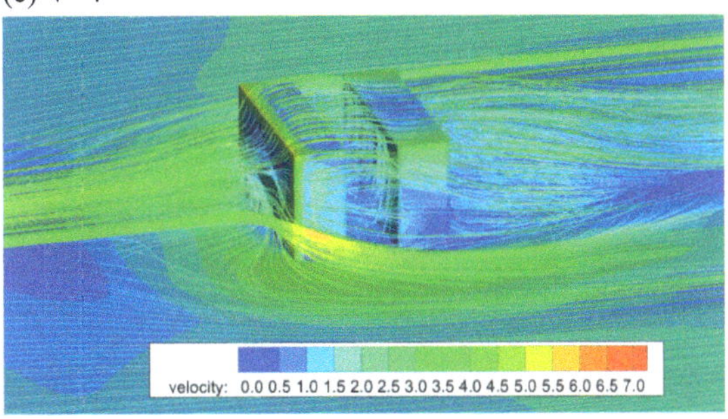

(d) V=5

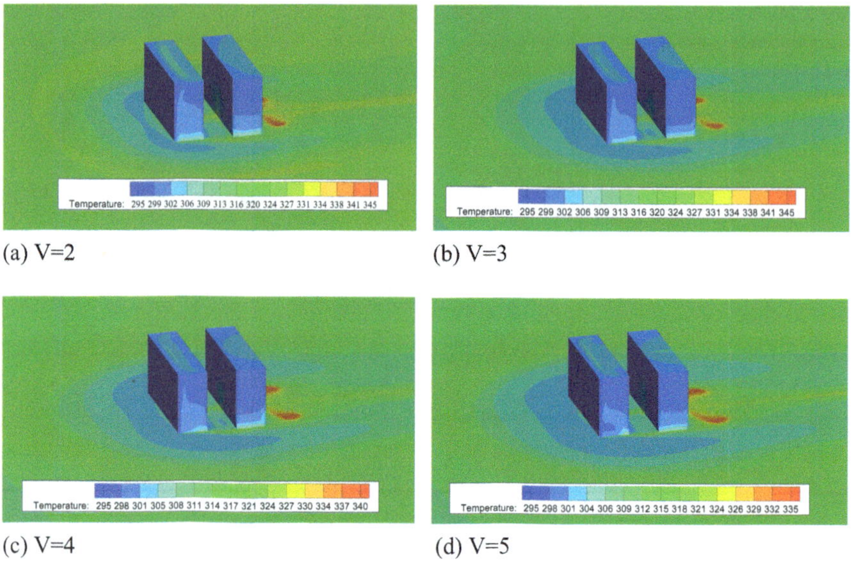

Fig. 2.25 Temperature distributions under different approaching wind speeds

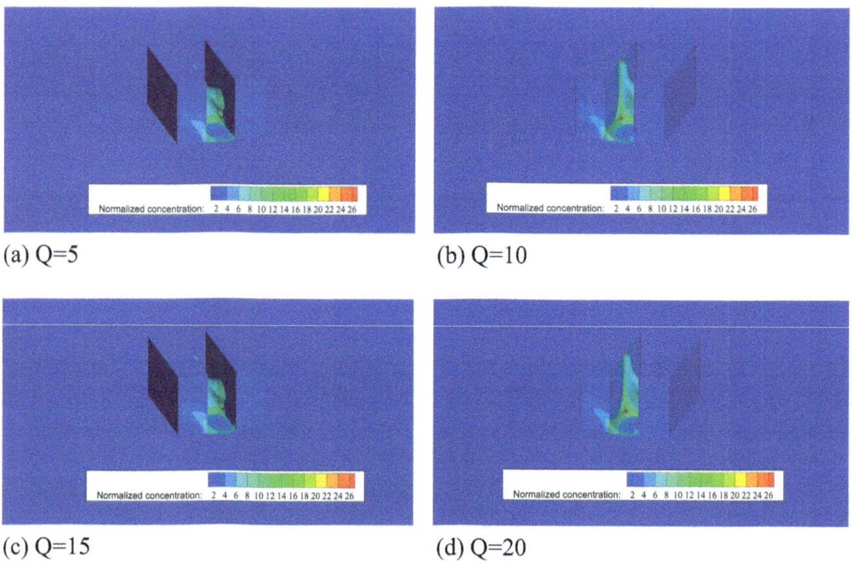

Fig. 2.26 Normalized concentration of pollutants under different source strength of pollutants

denotes the normalized concentration of pollutants under different source strength of pollutants. It can be seen that large number of pollutants accumulated in the middle of the street, and most of them near the leeward wall surface of the canyon

Fig. 2.27 Normalized concentration of pollutant under different positions of pollutant source

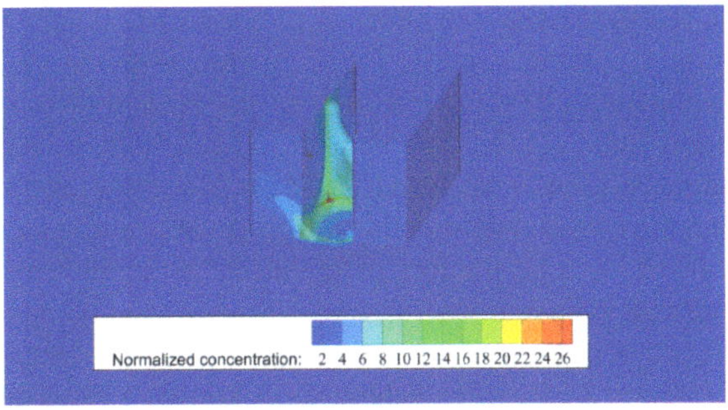

(a) The pollutant source at the street canyon

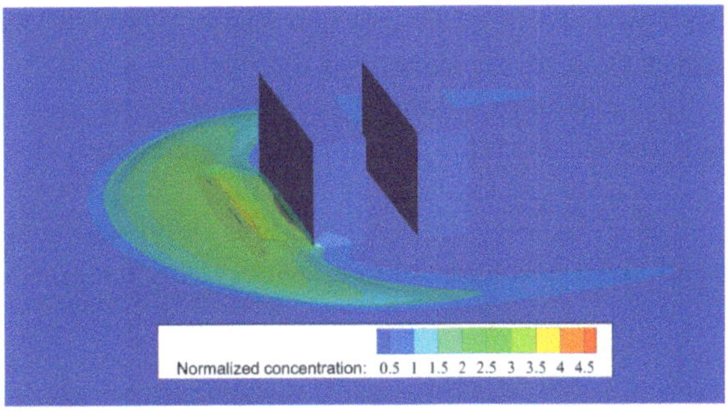

(b) The pollutant source at the windward

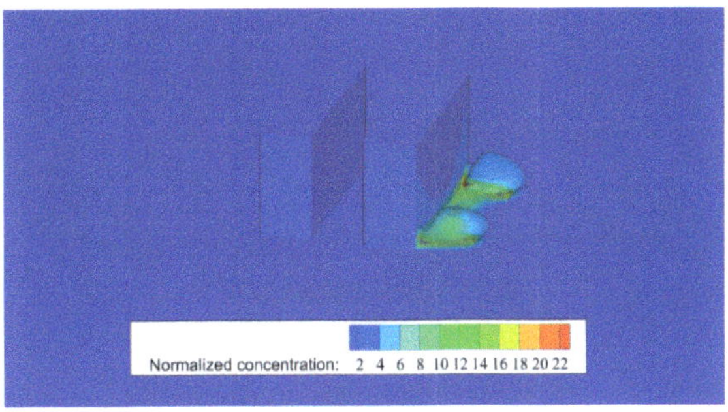

(c) The pollutant source at the leeward

which is similar to the temperature distribution. This is because of the vortex circulations produced by airflows in three directions, i.e., upward, downward and lateral. Meanwhile, it is necessary to mention that, the three-dimensional pollutant dispersion is different from that of the two-dimensional. In fact, contaminants can be transported in the opposite direction of the approaching wind due to the eddies and mean flows induced by buildings. Hence this behavior makes it becomes complex in three-dimensional pollutant dispersion. In order to accurately predict the pollutant dispersion, three-dimensional flow is necessary. On the other hand, the distributions of normalized concentration are almost same vary from Fig. 2.26a–d under different source strength of pollutants, which shows a good consistent with two-dimensional pollutant dispersion.

To investigate the effect of position of pollutant source on pollution dispersion within a street canyon, Fig. 2.27 demonstrates the normalized concentration of pollutant under three positions of pollutant source, i.e. at the street canyon, at the windward and at the leeward. The results revealed that the distributions of dimensionless normalized pollutant concentrations are influenced significantly by the three positions of pollutant source. When the pollutant source is set at the street canyon, a large number of pollutants accumulated in the middle of street that is difficult to be removed. In contrast, when the pollutant source is set at the windward direction of the buildings, relatively low-level pollutant concentrations is found that is diluted by the approaching wind. Besides, when the pollutant source is set at the leeward direction of the buildings, the pollutants also are difficult to diffuse. Thus the flow patterns of mixing and transport largely control the pollutant dispersion, which includes the shape and motion of vortex circulations within the flow field.

In the actual city, the streets are generally criss-cross. Urban designers would like to make the main streets parallel to the city's dominant wind direction, in order to maximize the use of natural ventilation in urban areas. Actually, the main street acts as a channel to the approaching wind where the streamwise velocity is relatively large in contrast to that in street canyons in which vortexes exist. Figure 2.28

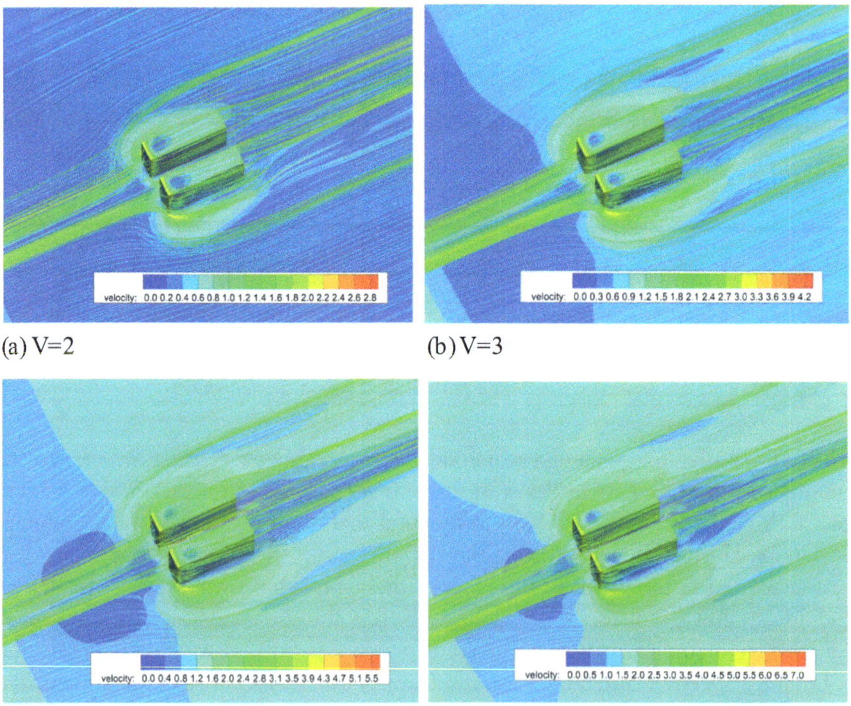

(a) V=2 (b) V=3

(c) V=4 (d) V=5

Fig. 2.28 3D streamlines under different approaching wind speeds

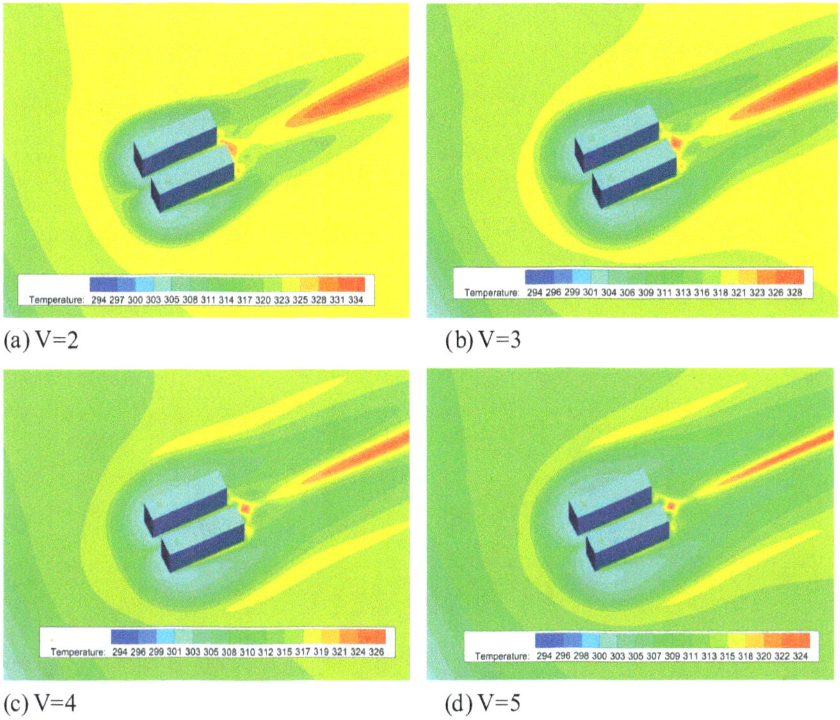

Fig. 2.29 Temperature distributions under different approaching wind speeds

displays the 3D streamlines under different approaching wind speeds for *H*/*B* = 2. It is obvious that the flow field has typical characteristics of bluff body aerodynamics. Separated flows in three directions are also formed when the approaching wind blocked by the wall surfaces of buildings. Vortex circulations are hardly found within the main street, which implies the drag force or blocking effect by buildings is negligible. Thus the air ventilation is supposed to be better in the street with the increase of approaching wind speed.

Figure 2.29 shows the temperature distributions under different approaching wind speeds when the main street parallel to the wind direction for *H*/*B* = 2. The results show that there is a slight decrease of mean temperature at the ground level with the increasing wind speeds. High level of temperature is found at the exit of the main street or some distance away it, because of the wake vortices and wake flows generated by the airflow behind the two buildings. Moreover, we find that the presence of buildings and street contribute to the high level of mean temperature at ground level when compared to that of the surrounding space. However, with the increase of approaching wind speed, the mean temperature of ground surface is to be lower and more uniform, which confirms that the increasing wind speed can significantly enhance the capacity of air ventilation in these studied cases.

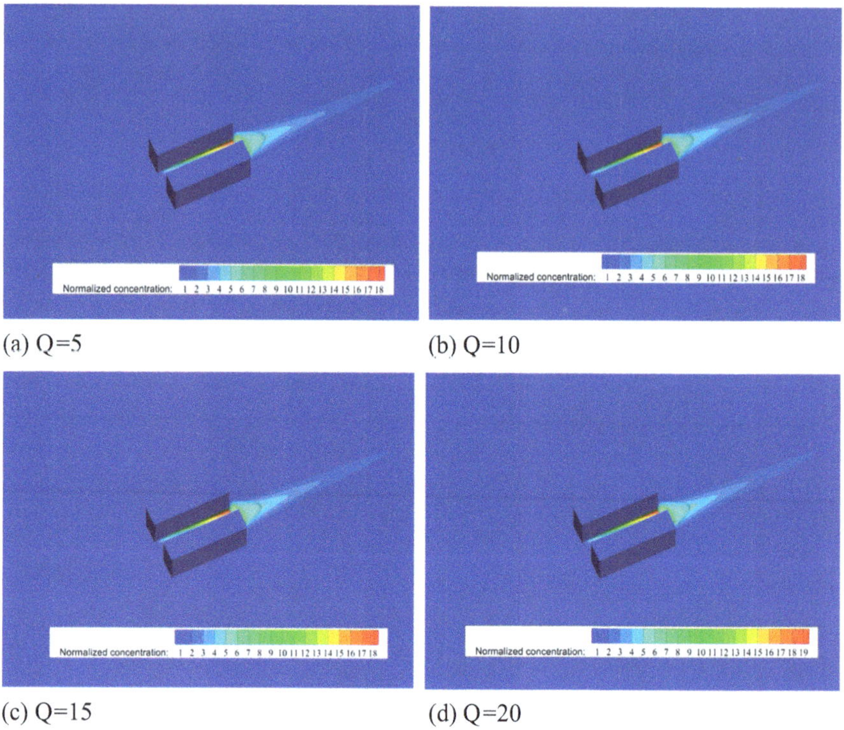

Fig. 2.30 Normalized concentration of pollutants under different source strength of pollutants

Once we analyzed the flow field and the temperature field of the main street between two buildings, it comes to the conclusion that the effect of approaching wind speed on air ventilation is significant. So there is no doubt that the pollutant concentrations will be less with the increase of wind speeds, but how's the pollutant distributions under different source strengths will be? To figure it out, we investigate the normalized concentration of pollutants under different source strength of pollutants, as shown in Fig. 2.30. It can be seen that not all contaminants can be blown away by the airflow within the main street, while some pollutant concentrations are accumulated at the leeward region of the main street. This can be attributed to the wake flows especially some wake vortices generated behind the buildings make the poor capacity for ventilation of local region. In addition, with the increase of source strength of pollutants, there is not an obvious change on the dimensionless normalized concentrations between these four studied cases. That is to say, the dispersion patterns of pollutants hardly changed with the different source strength of pollutants. But it is worth noting that, the normalized concentrations are inversely proportional to the calculated mean concentrations, thus the mean concentrations should increase with the increasing source strength of pollutants. As a result, the flow patterns and dispersion patterns within the main street are relatively simple compared to those of street canyon, the most direct reason is the streamwise velocity is relatively large and less vortexes exists, resulted in better air ventilation within the street.

2.8 Conclusions

A comprehensive mathematical model including the mass, momentum, and energy equations to describe the fluid flow and heat transfer characteristics, species transport equation to describe the pollutant transport mechanism, and RNG k–ε turbulence model to analyze the turbulent flow has been put forward. From the numerical simulation shown above, we can find:

(1) The street canyon with high-rise buildings may suppress the air ventilation, leads to a direct impact on human health. Recirculation flows induced by those vortexes result in poor capacity of dispersing ground-level pollutants, thus the traffic-related pollutants are difficult to be removed out through these deep canyons. Temperature distribution and the simulated pollutant concentrations are independent on approaching wind speed. In addition, source strength has little impact on the normalized dimensionless pollutant concentrations.
(2) The flow regime within the street canyons varies from multi-vortex regime to SF with the increase of street width. The turbulent mixing within the street canyons is stronger resulted in better air exchange and ventilation with the wider street width. With the increase of the street width, the temperature distribution within the street canyons becomes more uniform, the air ventilation within the canyon is better, thus less pollutant is accumulated.
(3) Air ventilation in front of the taller buildings may be better than those behind the taller buildings. When the taller buildings located behind the shorter buildings in the direction of approaching wind, the air exchange between the street canyons with external airflows is better. Arranging the taller buildings behind the shorter buildings may obtain relatively better air ventilation.
(4) The 3D flow characteristics are less dependent upon the approaching wind speed. The place in the windward direction of buildings with a low mean temperature at ground level, whereas the place in the leeward direction of buildings with a comparative high mean temperature at ground level. The flow patterns of mixing and transport largely control the pollutant dispersion, which includes the shape and motion of vortex circulations within the flow field. It also confirmed that make the street parallel to the city's dominant wind direction, the natural ventilation in urban areas is getting better.

References

1. J. Gallagher, L.W. Gill, A. Mcnabola. Numerical modelling of the passive control of air pollution in asymmetrical urban street canyons using refined mesh discretization schemes. Building & Environment. 56 (2012) 232–40.
2. T.R. Oke. Street design and urban canopy layer climate. Energy & Buildings. 11 (1988) 103–13.
3. J.J. Kim, J.J. Baik. A Numerical Study of Thermal Effects on Flow and Pollutant Dispersion in Urban Street Canyons. Journal of Applied Meteorology. 38 (1999) 1249–61.
4. S.M. Salim, R. Buccolieri, A. Chan, S.D. Sabatino. Numerical simulation of atmospheric pollutant dispersion in an urban street canyon: Comparison between RANS and LES. Journal of Wind Engineering & Industrial Aerodynamics. 99 (2011) 103–13.
5. Y. Tominaga, T. Stathopoulos. CFD simulation of near-field pollutant dispersion in the urban environment: A review of current modeling techniques. Atmospheric Environment. 79 (2013) 716–30.
6. J.F. Sini, S. Anquetin, P.G. Mestayer. Pollutant dispersion and thermal effects in urban street canyons. Atmospheric Environment. 30 (1996) 2659–77.
7. R.N. Meroney, M. Pavageau, S. Rafailidis, M. Schatzmann. Study of line source characteristics for 2-D physical modelling of pollutant dispersion in street canyons. Journal of Wind Engineering & Industrial Aerodynamics. 62 (1996) 37–56.
8. X. Xie, Z. Huang, J. Wang. The impact of urban street layout on local atmospheric environment. Building & Environment. 41 (2006) 1352–63.
9. X.X. Li, C.H. Liu, D. Leung. Numerical investigation of pollutant transport characteristics inside deep urban street canyons. Atmospheric Environment. 43 (2009) 2410–8.
10. J. Levitin, J. Haerkoenen, J. Kukkonen, J. Nikmo. Evaluation of the CALINE4 and CAR-FMI models against measurements near a major road. Atmospheric Environment. 39 (2005) 4439–52.
11. J. Berger, S.E. Walker, B. Denby, R. Berkowicz, P. Løfstrøm, M. Ketzel, et al. Evaluation and inter-comparison of open road line source models currently in use in the Nordic countries. Boreal Environment Research. 15 (2008) 319–34.
12. A. Venkatram, V. Isakov, R. Seila, R. Baldauf. Modeling the impacts of traffic emissions on air toxics concentrations near roadways. Atmospheric Environment. 43 (2009) 3191–9.
13. H. Chen, S. Bai, D.S. Eisinger, D. Niemeier, M. Claggett. Predicting Near-Road PM2.5 Concentrations: Comparative Assessment of CALINE4, CAL3QHC, and AERMOD2009.
14. C.C.C. Wong, C.H. Liu. Pollutant Plume Dispersion in the Atmospheric Boundary Layer over Idealized Urban Roughness. Boundary-Layer Meteorology. 147 (2013) 281–300.
15. T.L. Chan, G. Dong, C.W. Leung, C.S. Cheung, W.T. Hung. Validation of a two-dimensional pollutant dispersion model in an isolated street canyon. Atmospheric Environment. 36 (2002) 861–72.
16. C.H. Chang, J.S. Lin, C.M. Cheng, Y.S. Hong. Numerical simulations and wind tunnel studies of pollutant dispersion in the urban street canyons with different height arrangements. Journal of Marine Science & Technology. 21 (2013) 119–26.
17. Y.D. Huang, W.R. He, C.N. Kim. Impacts of shape and height of upstream roof on airflow and pollutant dispersion inside an urban street canyon. Environmental Science & Pollution Research International. 22 (2015) 2117–37.
18. Y. Huang, Z. Zhou. A Numerical Study of Airflow and Pollutant Dispersion Inside an Urban Street Canyon Containing an Elevated Expressway. Environmental Modeling & Assessment. 18 (2013) 105–14.
19. Y. Miao, S. Liu, Y. Zheng, S. Wang, Y. Li. Numerical Study of Traffic Pollutant Dispersion within Different Street Canyon Configurations. Advances in Meteorology. 2014 (2014) 1–14.
20. J.J. Kim, J.J. Baik. A numerical study of the effects of ambient wind direction on flow and dispersion in urban street canyons using the RNG $k-\varepsilon$ turbulence model. Atmospheric Environment. 38 (2004) 3039–48.

21. J. Hang, M. Sandberg, Y. Li, L. Claesson. Pollutant dispersion in idealized city models with different urban morphologies. Atmospheric Environment. 43 (2009) 6011–25.
22. Y. Huang, X. Xu, Z.Y. Liu, C.N. Kim. Effects of Strength and Position of Pollutant Source on Pollutant Dispersion Within an Urban Street Canyon. Environmental Forensics. 16 (2015) 163–72.
23. S.J. Mei, C.W. Liu, L. Di, F.Y. Zhao, H.Q. Wang, X.H. Li. Fluid mechanical dispersion of airborne pollutants inside urban street canyons subjecting to multi-component ventilation and unstable thermal stratifications. Science of the Total Environment. 565 (2016) 1102–15.
24. G. Mills. An urban canopy-layer climate model. Theoretical & Applied Climatology. 57 (1997) 229–44.
25. N.Y.O. Tong, D.Y.C. Leung. Effects of building aspect ratio, diurnal heating scenario, and wind speed on reactive pollutant dispersion in urban street canyons. Journal of Environmental Sciences. 24 (2012) 2091–103.
26. P. Wang, D. Zhao, W. Wang, H. Mu, G. Cai, C. Liao. Thermal Effect on Pollutant Dispersion in an Urban Street Canyon. International Journal of Environmental Research. 5 (2011) 813–20.
27. B. Pulvirenti, S.D. Sabatino, R. Buccolieri. Thermal effects on flow and pollutant dispersion within street canyons. Turbulence, Heat and Mass Transfer 6 Proceedings of the Sixth International Symposium On Turbulence, Heat and Mass Transfer 2009. pp. E-0210.
28. D.M.S. Madalozzo, A.L. Braun, A.M. Awruch, I.B. Morsch. Numerical simulation of pollutant dispersion in street canyons: Geometric and thermal effects. Applied Mathematical Modelling. 38 (2014) 5883–909.
29. W.W. Li, R.N. Meroney. Gas dispersion near a cubical model building. Part I. Mean concentration measurements. Journal of Wind Engineering & Industrial Aerodynamics. 12 (1983) 15–33.
30. Y. Ogawa, S. Oikawa, K. Uehara. Field and wind tunnel study of the flow and diffusion around a model cube—II. Nearfield and cube surface flow and concentration patterns. Atmospheric Environment. 17 (1983) 1161–71.
31. R.N. Meroney. Bluff-body aerodynamics influence on transport and diffusion. Journal of Wind Engineering & Industrial Aerodynamics. 33 (1990) 21–33.
32. J. Hang, Y. Li, M. Sandberg, R. Buccolieri, S.D. Sabatino. The influence of building height variability on pollutant dispersion and pedestrian ventilation in idealized high-rise urban areas. Building & Environment. 56 (2012) 346–60.
33. J. Hang, Y. Li, M. Sandberg. Experimental and numerical studies of flows through and within high-rise building arrays and their link to ventilation strategy. Journal of Wind Engineering & Industrial Aerodynamics. 99 (2011) 1036–55.
34. M.F. Yassin, R. Kellnerová, Z. Janour. Numerical simulation on pollutant dispersion from vehicle exhaust in street configurations. Environmental Monitoring & Assessment. 156 (2009) 257–73.
35. S.J. Park, J.J. Kim, M.J. Kim, R.J. Park, H.B. Cheong. Characteristics of flow and reactive pollutant dispersion in urban street canyons. Atmospheric Environment. 108 (2015) 20–31.
36. L. Yang, Y. Li. Thermal conditions and ventilation in an ideal city model of Hong Kong. Energy & Buildings. 43 (2011) 1139–48.
37. Y. Miao, S. Liu, Y. Zheng, S. Wang, Z. Liu, B. Zhang. Numerical study of the effects of Planetary Boundary Layer structure on the pollutant dispersion within built-up areas. Journal of Environmental Sciences. 32 (2015) 168–79.
38. Peng, Ming, Cheng, Wu. Modeling Thermal Comfort and Optimizing Local Renewal Strategies—A Case Study of Dazhimen Neighborhood in Wuhan City. Sustainability. 7 (2015) 3109–28.
39. C. Peng, T. Ming, J. Gui, Y. Tao, Z. Peng. Numerical analysis on the thermal environment of an old city district during urban renewal. Energy & Buildings. 89 (2015) 18–31.
40. Y. Tominaga, A. Mochida, R. Yoshie, H. Kataoka, T. Nozu, M. Yoshikawa, et al. AIJ guidelines for practical applications of CFD to pedestrian wind environment around buildings. Journal of Wind Engineering & Industrial Aerodynamics. 96 (2008) 1749–61.

41. FlunetInc. Fluent user's guide. 2006.
42. F. Kreith. Principles Of Sustainable Energy Systems, Second Edition. Crc Press. (2013).
43. K. Kimura, D.G. Stephenson. Solar radiation on cloudy days. Ashrae Journal. (1969).
44. K. Uehara, S. Murakami, S. Oikawa, S. Wakamatsu. Wind tunnel experiments on how thermal stratification affects flow in and above urban street canyons. Atmospheric Environment. 34 (2000) 1553–62.

Chapter 3
Thermal Comfort and Pollutant Transport in Outdoor Street Buildings

3.1 Introduction

With the development of economic construction in China, the scale of city is on a high speed extension [1, 2]. Then various kinds of buildings are being erected in these cities, which leads to the drastic change of urban surface structures [3–7] and properties thereof [8, 9] (Fig. 3.1). The area proportion of cement, asphalt road surface, and building of concrete structure keeps growing, whereas that of greenbelt and water body decreases all the way [10, 11]. With soil, water level, and the coverage type of underlying surface tapering off, the outspread of impermeable surface composed of cement, asphalt, and metal mixtures has caused a series of problems, such as the weakening of water evaporation on the earth's surface [12], the acceleration of runoff [13], the increase of sensible heat storage and transfer [14], and the water quality deterioration [15]. Meanwhile, the sharp decrease of greenbelts, water bodies, and forests leads to the insufficient absorption of greenhouse gases, which makes the city be a heat source and/or a huge heat island [16]. Additionally, the drastic increase of urban population density also leads to the sharp increase of anthropogenic discharge, due to which the urban heat island effect has gradually become an important factor influencing the urban living environment, residents' health, and the urban and regional climate [17, 18].

For another, the economic and social activities of high population density will inevitably lead to large emission of particular matter (PM 2.5) which will, once beyond the capability and the load of atmospheric circulation, cause the accumulation therefrom, and lead to the large scale of haze [19, 20]. In 2013, "haze" became one of the annual keywords. In January of this year, haze covered over 30 provinces (cities and districts) [21]. In Beijing, only 5 days of the month had blue sky [22]. A recent survey shows that less than 1% of the 500 metropolises in China 500 have reached the air quality standards recommended by World Health

T. Ming et al., *Pollutant Dispersion in Built Environment*,
DOI 10.1007/978-981-10-3821-1_3

Fig. 3.1 A glimpse of Wuhan

Organization. On the other hand, for the top 10 most seriously polluted cities around the world, seven are in China. In January 4, 2014, China National Commission for Disaster Reduction and China Ministry of Civil Affairs took the hazardous haze into the Notification of Natural Hazard 2013 for the first time. In February 2014, the top priority of handling haze and improving the air quality is to control PM 2.5, which should be realized from the perspective of reducing the coal consumption and vehicle emission, adjusting industry, strengthening management, establishing joint control and prevention, and governance according to the law, focusing on important areas, servicing key performance of indication, enhancing environmental law-enforcing supervision, and investigating responsibilities therefrom.

In this work, a numerical simulation was conducted using PHOENICS so as to assess the wind thermal environment of an urban area according to its current city structure, architectural composition, materials of ground surface, distribution of pollution source, environmental conditions, and solar radiation. The thermal environment and pollutant transport mechanism were analyzed, so were the problems and influential characteristics of current city layout.

3.2 The Description of Model

3.2.1 The Physical Model

First of all, a geometrical model including the architectures, greenbelt, land, territory, and water body was established via SketchUp according to the actual plan, as shown in Fig. 3.2. Afterward, the architectural models were imported into PHOENICS to conduct a three-dimensional flow numerical simulative calculation. The total length of the model is 458 m, the width is 496 m, and the height is 100 m.

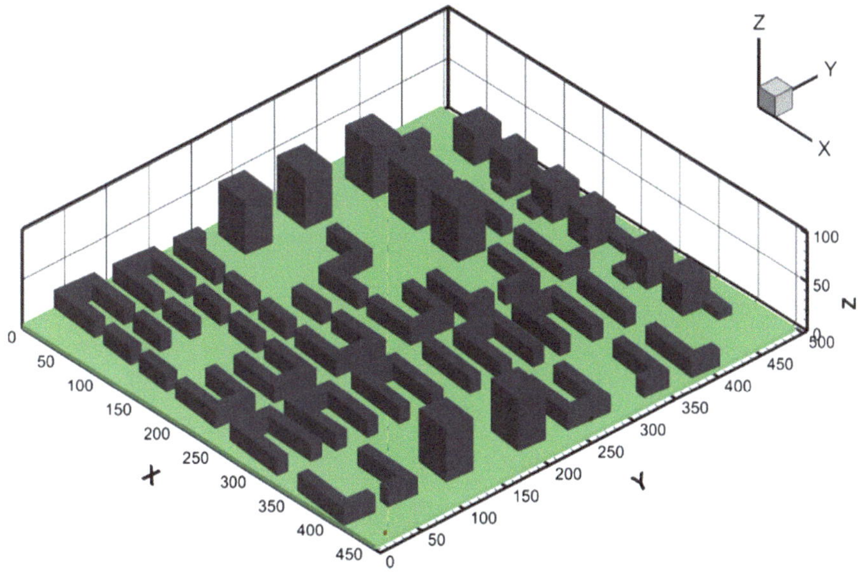

Fig. 3.2 The geometrical model from street scale

3.2.2 Mathematical Model and Numerical Method

The airflow in the atmospheric convective boundary layer usually belongs to incompressible and low-speed turbulence, which is calculated by the method of large eddy simulation (LES) and standard $k-\varepsilon$ turbulent model. On applying LES to large-scale model, the main difficulties are high requirement of numerical schemes and huge computational cost, so standard $k-\varepsilon$ turbulent with less fluctuation and high accuracy is applied to numerically simulate airflow. The natural flow should be vertically upward due to the effect of horizontal ground surface having high temperature, and flow regimes can also be classified according to the ratio of Gr and Re^2 shown as

$$\frac{Gr}{Re^2} = \frac{(g\beta\Delta TL^3/v^2)}{(UL/v)^2} = \frac{g\beta\Delta TL}{U^2}, \tag{3.1}$$

where Gr and Re are the Grashof and Reynolds numbers, respectively; ΔT is the maximum temperature increase within the area; v, β, and L are, respectively, the air viscosity, thermal expansion coefficient, and the maximum building height; g is the acceleration of gravity; and U is the average wind velocity in the area.

As a result, mass equation, Navier–Stokes equation, energy equation, and standard k–ε equations can be written as follows:

Mass equation:

$$\frac{\partial \rho}{\partial t} + \frac{\partial (\rho u_i)}{\partial x_i} = 0. \tag{3.2}$$

Navier–Stokes equation:

$$\frac{\partial (\rho u_i)}{\partial x_i} + \frac{\partial (\rho u_i u_j)}{\partial x_j} = \rho g_i - \frac{\partial p}{\partial x_i} + \frac{\partial \tau_{ij}}{\partial x_j}. \tag{3.3}$$

Energy equation:

$$\frac{\partial (\rho c_p T)}{\partial t} + \frac{\partial (\rho c_p u_j T)}{\partial x_j} = \frac{\partial}{\partial x_j}\left(\lambda \frac{\partial T}{\partial x_j}\right) + \tau_{ij} + \frac{\partial u_i}{\partial x_j} + \beta T\left(\frac{\partial p}{\partial t} + u_j \frac{\partial p}{\partial x_j}\right). \tag{3.4}$$

Equation for the turbulent kinetic energy k

$$\frac{\partial (\rho k)}{\partial t} + \frac{\partial (\rho k u_i)}{\partial x_j} = \frac{\partial}{\partial x_j}\left(\left(\mu + \frac{\mu_t}{\sigma_k}\right)\frac{\partial k}{\partial x_j}\right) + G_k - \rho \varepsilon. \tag{3.5}$$

Equation for the turbulent kinetic energy dissipation rate ε:

$$\frac{\partial (\rho \varepsilon)}{\partial t} + \frac{\partial (\rho k u_j)}{\partial x_j} = \frac{\partial}{\partial x_j}\left(\left(\mu + \frac{\mu_t}{\sigma_\varepsilon}\right)\frac{\partial \varepsilon}{\partial x_j}\right) + C_{1\varepsilon}\frac{\varepsilon}{k}G_k - C_{2\varepsilon}\rho\frac{\varepsilon^2}{k}. \tag{3.6}$$

In the equations above, the variables such as the velocities and temperatures are all time averaged values, and the stress tensor *ij* is given by

$$\tau_{ij} = \eta \frac{\partial u_i}{\partial x_j} - \rho \overline{u_i' u_j'}. \tag{3.7}$$

G_k denotes the generation of turbulence kinetic energy because of the mean velocity gradient; $C_{1\varepsilon}$ and $C_{2\varepsilon}$ are the empirical constants; σ_ε and σ_k represent the turbulent Prandtl numbers for ε and *k*, respectively; and μ_t is the eddy viscosity:

$$G_k = \mu_t\left(\frac{\partial u_i}{\partial x_j} + \frac{\partial u_j}{\partial x_i}\right)\frac{\partial u_i}{\partial x_j} \tag{3.8}$$

$$\mu_t = \rho C_\mu \frac{k^2}{\varepsilon}. \tag{3.9}$$

$C_{1\varepsilon} = 1.44$, $C_{2\varepsilon} = 1.92$, $C_\mu = 0.09$, $\sigma_\varepsilon = 1.3$ and $\sigma_k = 1.0$ are constant for turbulent model. Via the calculus of differences in PHOENICS, the second-order upwind difference scheme was adopted to guarantee the accuracy of calculation. The SIMPLE calculation in PHOENICS was adopted in the solution of pressure

Fig. 3.3 The mesh system of the geometrical model

and velocity coupling. To guarantee the mesh quality, PARSOL model was adopted, in addition with block encryption, to conduct the division structurally. The total mesh number is 250 × 250 × 47; hence 2.94 million in total (Fig. 3.3).

3.2.3 Boundary Conditions

In this study the most frequent wind direction of the model is southeast, with the mean velocity 2.6 m/s, and the maximum velocity 10 m/s. In the atmospheric boundary, the mean velocity of wind varies with height. Taking the unevenness of town, we adopted the exponential distribution of experience to describe the variation of mean velocity of wind near the surface layer with that of height, with the coefficient 0.2. According to the parameters of underlying surface and the local climatic data collected, the boundary conditions set in the calculation are presented in Table 3.1.

3.2.4 Simulation Assessment

During the planning of central urban area, the wind environment in building groups is the problem attracted broad attention, whose representation includes flowing, wind velocity distribution, pressure distribution, etc. From this perspective, beside the overall wind field distribution in the central urban area, we should pay more

Table 3.1 Boundary conditions for the calculation model

The parameters of underlying surface					
Boundary type		Source of energy	Thermo conduction coefficient	The absorption coefficient of solar radiation	Emissivity of radiation
Architectures		Adiabatic	10	0.2	0.9
Water body		Fixed temperature	Thermal insulation	0.96	0.96
Field		Adiabatic	30	0.8	0.4
Road		Adiabatic	30	0.57	0.93
Meteorological parameters					
	Wind velocity (m/s)	Wind direction	The height of wind velocity measuring	Outdoor temperature (°C)	Direct solar radiation and scattering intensity (W/m^2)
Summer	2.6	SE	10	26.5	700,609
Winter	2.6	SE	10	0.48	350.112

attention to the location of vortex, wind shade, (windless region) and blast area, the location where the wind velocity varies sharply, and the inappropriate location due to the pressure difference between the windward side and lee side. These factors could directly portray the wind environment of architectures in the urban area. At the meantime, the air flow would directly influence the heat convection between skin and environment and the water evaporation on the surface of skin. There are many factors influencing the air flow, such as the velocity of air stream (speed rate), angle, turbulence intensity, and frequency. However, precedent researches show that some factors such as the angle of stream have little impact on the comfortable degree of human physical activities. Therefore, among those factors in composition of thermal environment, only the wind velocity parameters influencing the human thermal sensation are taken into account.

Thermal environment refers to the one composed via those factors such as solar radiation, temperature, surface temperature of surrounding objects, relative humidity, and air stream velocity, which acts on and influences the human heat sensation and health. Solar radiation directly influences the weather conditions near the earth's surface, which makes great efforts on the local hygrothermal environment, wind environment, and atmospheric environment of cities and functional areas, as well as has much importance with the creation of favorable urban physical environment and the correct recognition of solar radiation. There are many factors which influence the thermal environment of central urban areas. Those that could be sensed and visualized include temperature, humidity, solar radiation, and virescence statues. Temperature is one of the important parameters that could be felt and expressed by human beings, which could reflect the influence of solar radiation received, the humidity, the wind velocity, the virescence, and the water surface on the thermal environment of urban area. Meanwhile, since the surrounding thermal environment is also an important factor closely relative with human being's living quality, the assessment of thermal environment of central urban area is also necessary.

3.3 Block Dimension

3.3.1 *The Wind Thermal Working Conditions in Summer*

Figure 3.4 indicates the wind velocity distribution of M Street of A County under the condition that the wind direction is SE, the velocity is 2.6 m/s, and the height is 1.5 m. Since the platy buildings in the southeast are perpendicular to the incoming wind, the impact on the building forms the undershoot wind which flows back near the ground, forming inconspicuous vortex on the weather side. When the incoming wind penetrates into the district via the channel inter-buildings, the wind tunnel narrowed sharply, and the wind velocity increases; it activates the "funnelling" effect, with the maximum wind velocity 4 m/s. Meanwhile, at the corner of the building, the wind is accelerated during its bypass. Generally, from the perspective of the whole district, the wind velocity of main pedestrian areas is controlled within the range 0.5–5.0 m/s. Besides, there are no evident dead zones or vortex around buildings, which is conductive to the natural ventilation in summer. At the corner of the weather side, the local wind is accelerated. It is suggested that landscape plants such as boarder trees should be planted to act as the screen to reduce the wind velocity in the pedestrian zone.

Figure 3.5 indicates the wind age of M Street of A County under the condition that the wind direction is SE, the velocity is 2.6 m/s, and the height is 1.5 m. Air age is an index of characteristic to indicate the indoor ventilation introduced by M. Sandberg in Sweden. It refers to the time consumption that the fresh air flows indoor and reaches certain point from the air intake. As a ventilation characteristic and an index to assess the ventilation efficiency, air age now has been widely applied in many fields.

It could be seen that the oldest air in the district ranges between 241 and 260 s, mainly distributed at the northwest. Meanwhile, in the center, the building group is of the half surrounded pattern, which generates vortexes, making the air refreshment much slower. This may lead to the massive gathering of pollutant. At the meantime, the air circulation around buildings becomes slower, the amount of fresh air is insufficient, and the air quality is low, which has a negative influence on the outdoor activities. The air age of greatest proportion ranges from 93 to 186 s, distributed in the middle and downstream area on the prevailing wind direction in the district. The air refreshment thereof is fast.

Figure 3.6 indicates the pressure distribution of M Street of A County under the condition that the wind direction is SE, the velocity is 2.6 m/s, and the height is 1.5 m. Since the platy buildings in the south will block the wind, the air flow on the weather side is therefore hindered, with its velocity reduced. Some kinetic energy of the wind has been transformed into static pressure, which increases the pressure on the weather side, forming positive pressure, whereas at the backside and bilateral sides of the building, the detour flow makes the air thinner and generates negative

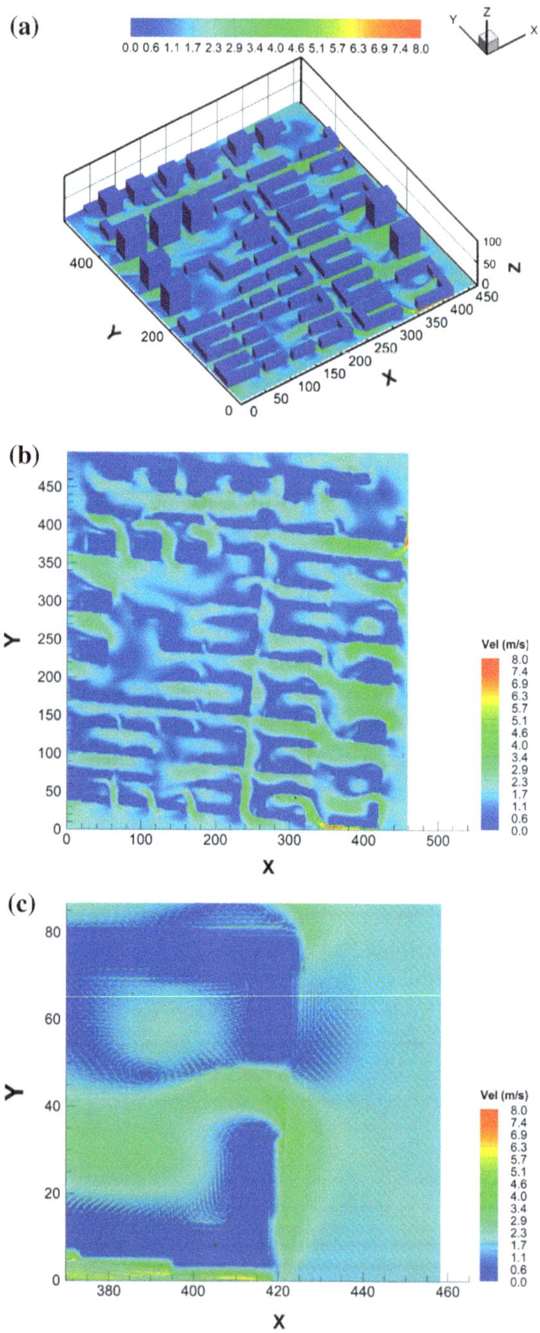

Fig. 3.4 Velocity distribution at the height of 1.5 m

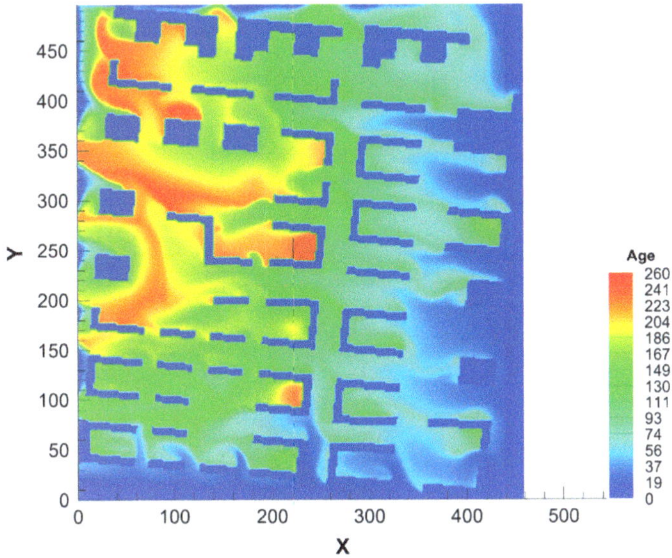

Fig. 3.5 Air age distribution at the height of 1.5 m (unit: s)

pressure. As shown in the figure, local high pressure appears at the southeast, with the value up to 10 Pa. The reason is that the wind velocity in this area is high. Buildings, especially those high-rise buildings, have significant blocking effect on the incoming wind. At the same time, different negative pressure zones appear at the backside, hence the northwest of the building, with −3 Pa the value.

Figure 3.7 indicates the predicted percentage dissatisfied revised (PPDR) distribution of M Street of A County under the condition that the wind direction is SE, the velocity is 2.6 m/s, and the height is 1.5 m. The distribution of PPDR at 1.5 m the pedestrian height in summer is indicated in Fig. 3.7. Generally, the wind could be felt in the district. The PPDR ranks above 60%, hence, cool in summer; and however, that around some buildings is relatively lower, around 40%.

Figure 3.8 indicates the temperature distribution of M Street of A County under the condition that the wind direction is SE, the velocity is 2.6 m/s, and the height is 1.5 m. The simulation result of temperature in the whole thermal environment could, in the general sense, be adopted to assess the local thermal environment status whose significance is to derive the hot island value therefrom. The temperature simulation on the surface of buildings could be adopted to obtain the thermal environment for buildings, which could perfect the indoor thermal environment via building evaluation measures, and reduce the energy consumption generated from temperature increase.

The simulation results show that the mean temperature at the height of pedestrian, 1.5 m, is 28.9 °C. No large area of hot island is found, which means the ventilation of the district is good, with no excessive local temperature that may lead to the thermal discomfort of pedestrians.

The maximum temperature appears at the west of the district, namely the downstream of southeaster. The building materials have both strong absorption and

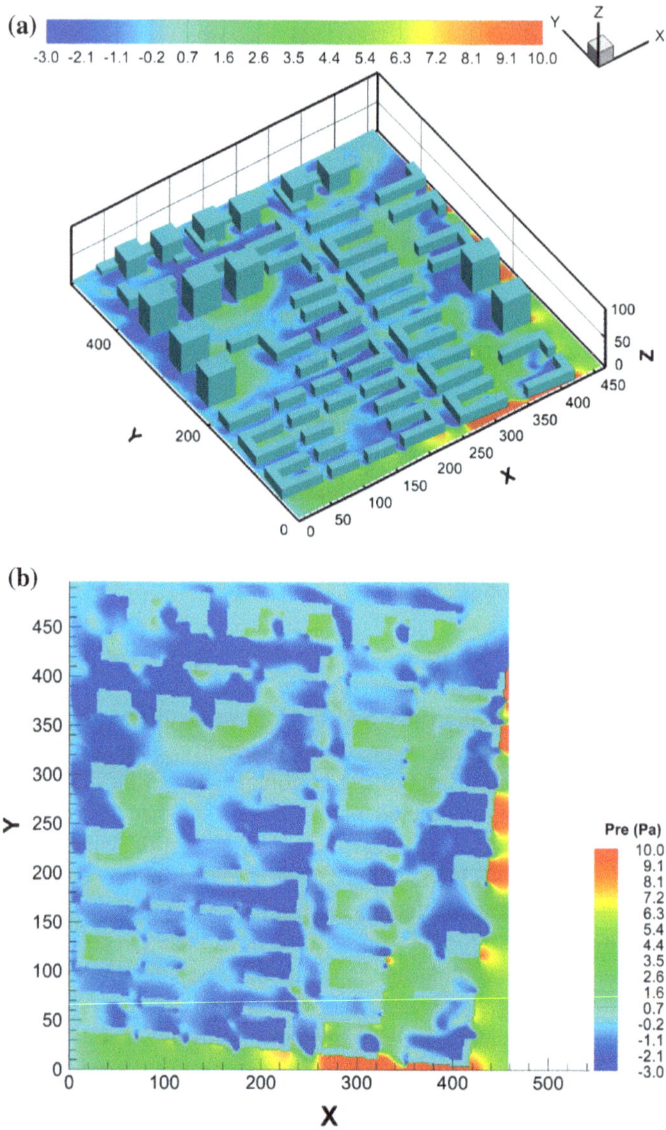

Fig. 3.6 Pressure distribution at the height of 1.5 m

emission of the solar radiation. When the southeaster penetrates the district, it will have heat exchange with the surface of the building, where the air is heated, making the temperature there on an increase. Therefore, the temperature downstream is higher than that upstream.

Figure 3.9 indicates the pollutant distribution of M Street of A County under the condition that the wind direction is SE and the velocity is 2.6 m/s. As shown in the

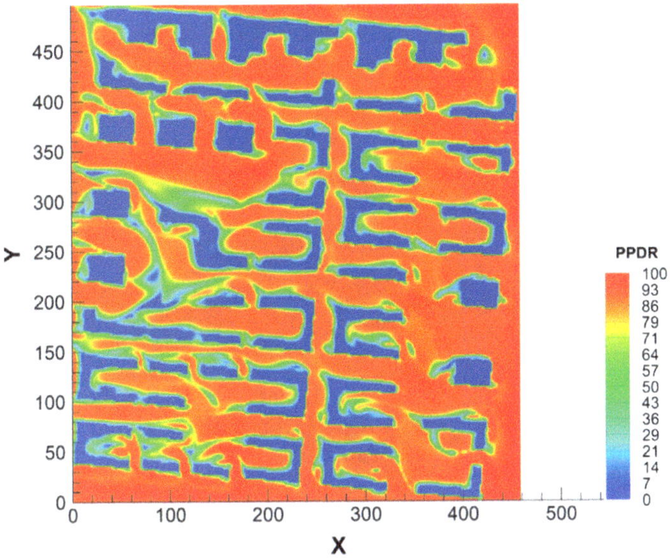

Fig. 3.7 PPDR distribution at the height of 1.5 m

figure, the pollutant dispersion in summer is similar to that in winter, which is maintained at certain height, no more than 50 m. Under the force of SE wind field, the pollutant on the *X*-axis disperses toward west, with a large dispersion scale, whereas on the *Y*-axis, the pollutant disperses toward north, whose concentration decreases with the increase of distance. Generally speaking, at the point $Y = 40$ where the wind velocity is insufficient and near the source of pollution, the pollutant climbing is not evident, which leads to the local pollutant gathering. At the points $Y = 60$ m and $Y = 100$ m, since the concentration is low, and the wind velocity is high, the pollutant dispersion toward downstream is evident. However, the influence is limited, which has nearly no impact about 350 m from the pollutant.

3.3.2 The Wind Thermal Working Conditions in Winter

Figure 3.10 indicates the wind velocity distribution of M Street of A County under the condition that the wind direction is SE, the velocity is 2.6 m/s, and the height is 1.5 m. Generally, the wind velocity at southeast is relative higher, with 4 m/s the mean value. In the center of district, the wind shade is evident. The wind velocity is no more than 1 m/s. The local wind velocity around the building group is high. This is because the detour flow will occur when some wind impacts the building, which will get superimposed with the incoming wind and forms the wind of greater velocity. This is called vortex phenomenon. The reason that the wind velocity is too low is that the block effect of buildings on wind is evident. The surrounding

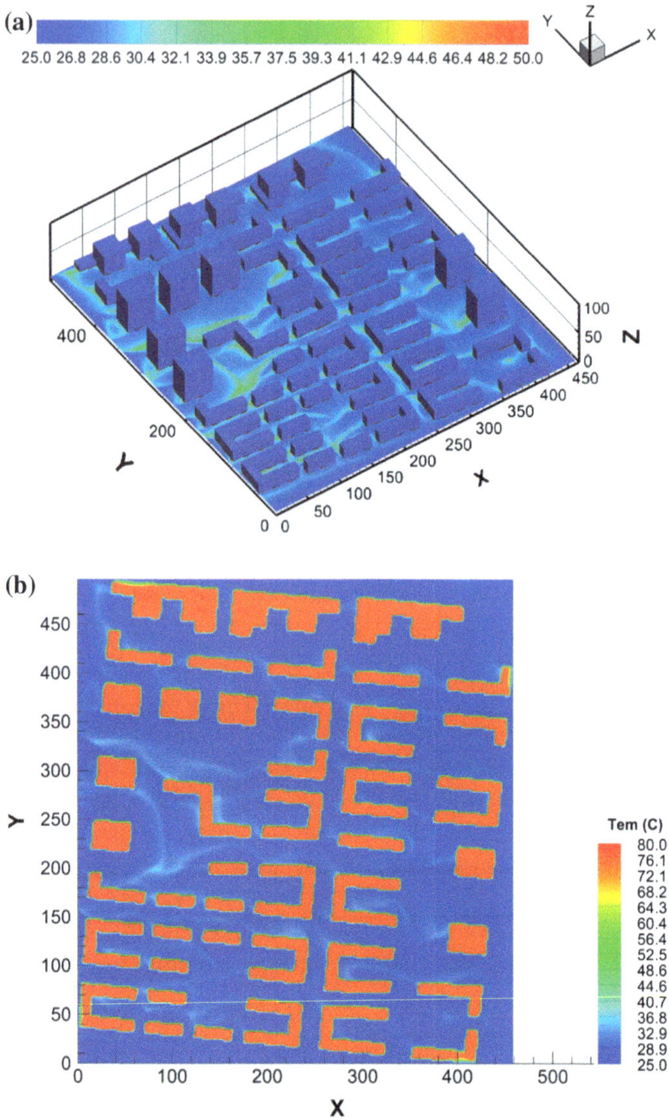

Fig. 3.8 Temperature distribution (the height is 1.5 m)

buildings block the wind completely, which reduces the velocity of some wind down to nearly 0. If so, the ventilation in the center of the district is poor, which might increase the local temperature. Therefore, the local ventilation of these areas should be intensified via some measures. In the scheme optimization, if the construction density in these local areas could be lowered, the ventilation shall be improved drastically, which is significantly conductive to the heat exchange.

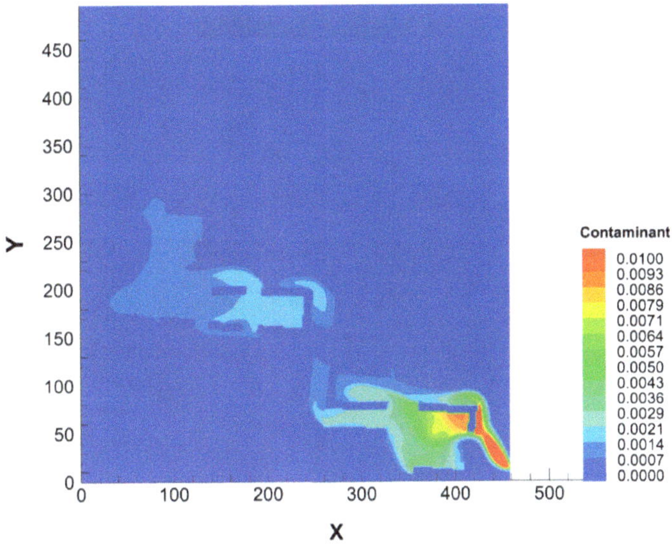

(**a**) The overall distribution of pollutant dispersion

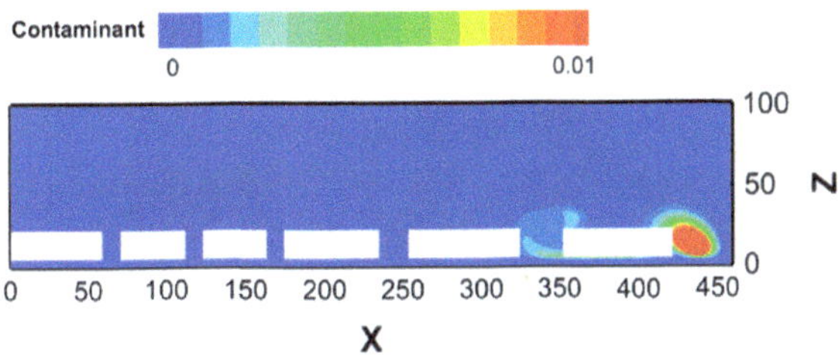

(**b**) The cross section of pollutant distribution at the point Y=40

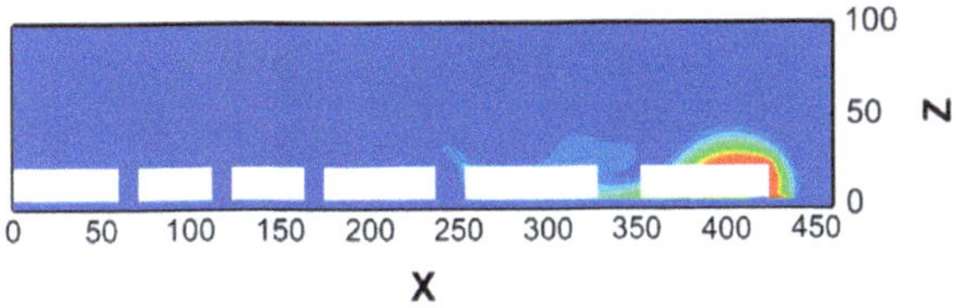

(**c**) The cross section of pollutant distribution at the point Y=60

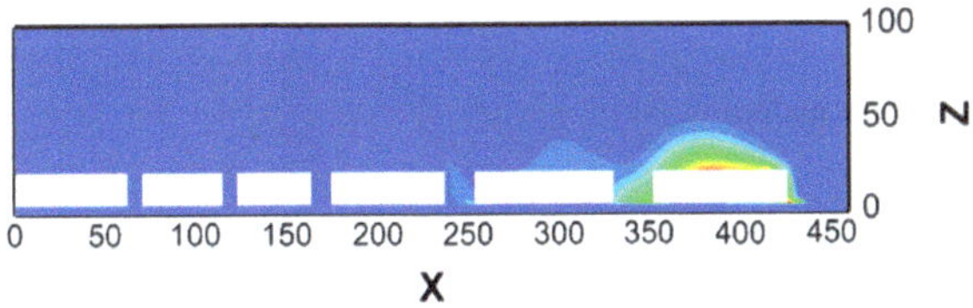

(**d**) The cross section of pollutant distribution at the point Y=80

Fig. 3.9 Pollutant distribution at different cross sections

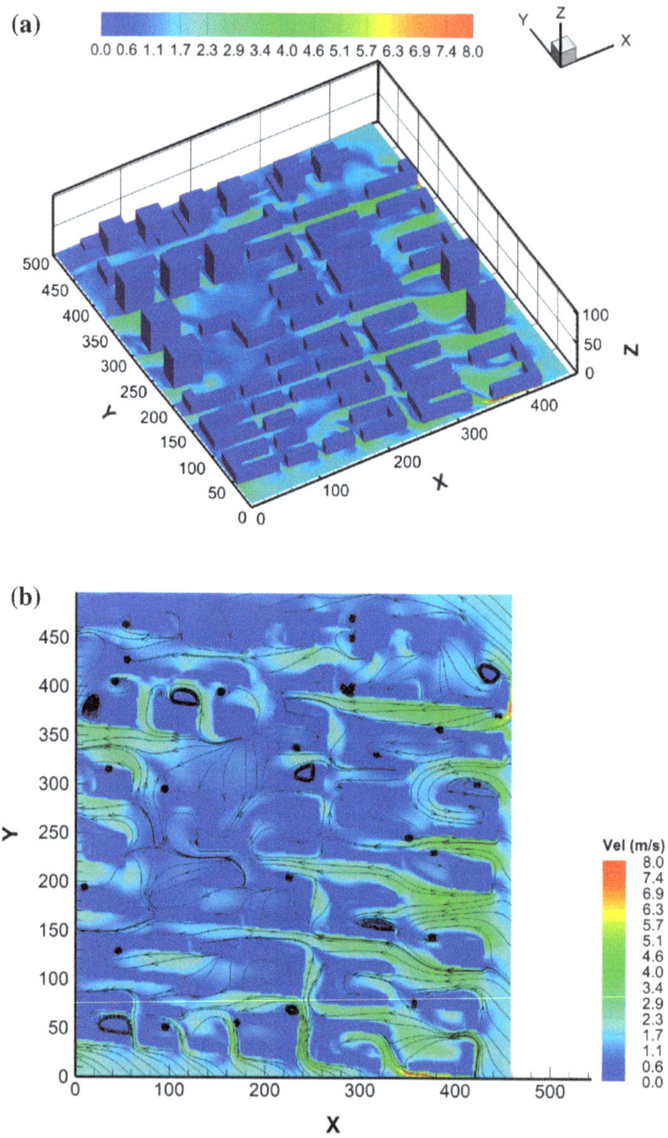

Fig. 3.10 Velocity distribution at the height of 1.5 m

Figure 3.11 indicates the wind air distribution of M Street of A County under the condition that the wind direction is SE, the velocity is 2.6 m/s, and the height is 1.5 m. As shown in the figure, in the area up wind, hence the southeast of the district, the air age is relatively younger, with more rapid pollutant dispersion. However, due to the arrangement of some buildings, the wind velocity is inclined to form dead angle in the district, hence wind shade. Especially those buildings whose

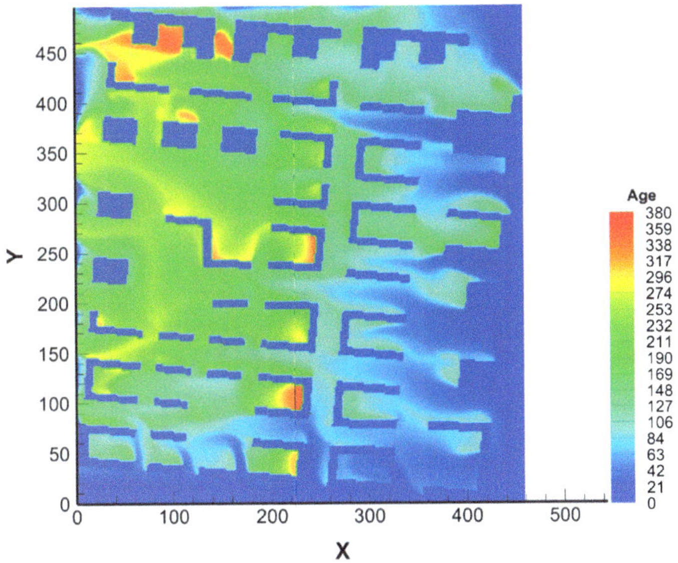

Fig. 3.11 Air age distribution at the height of 1.5 m

opening is toward the west have little air entered. The ventilation thus is poor. In the downwind area, hence the northwest of the district, the air age on average is relatively older than that of up wind. This is because the buildings on the up wind direction have blocked some incoming wind, which reduces the quantity of the air entering into the area and weakens the ventilation. On the other hand, the building arrangement has a strong influence on the air quantity; therefore, the optimization of which could increase the wind velocity, improving the ventilation in the district.

Figure 3.12 indicates the pressure distribution of M Street of A County under the condition that the wind direction is SE, the velocity is 2.6 m/s, and the height is 1.5 m. As shown in the figure, the local high pressure appears at the southeast of the district, with the value up to 10 Pa. The reason is that the wind velocity in these areas is high. The buildings especially high-rise buildings have significant blocking effect on the incoming wind. Meanwhile, different negative pressure zones appear at the northwest, with −3 Pa the value. It could be seen that the higher the buildings are, the higher the building density is, and the larger the negative pressure zone will be. As thus, due to the pressure difference between the front and the back of buildings, the wind velocity in these areas is very high. Furthermore, in the center of the district, the pressure has a homogeneous distribution, due to which the negative pressure zone is relatively smaller, with inconspicuous pressure difference. Therefore, the local wind velocity in these areas is relatively lower, which generates large area of wind shade. At the northwest of the district, little negative pressure zones appear at the back of high-rise buildings. Generally speaking, the height and the density of buildings have significant influence on the pressure distribution.

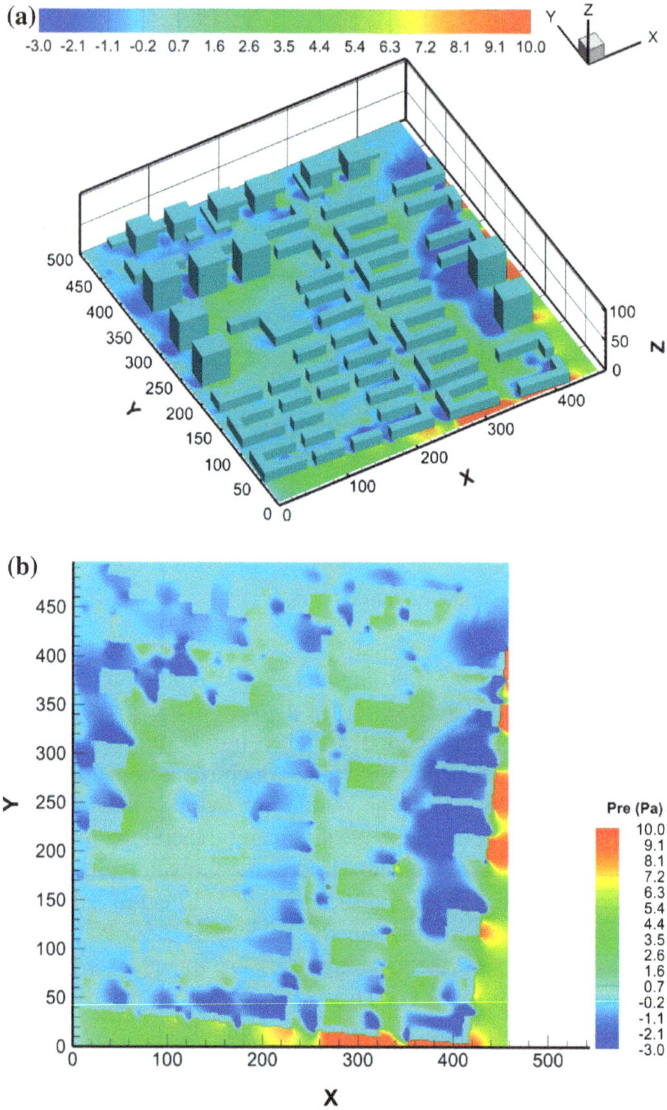

Fig. 3.12 Pressure distribution at the height of 1.5 m

Figure 3.13 indicates the temperature distribution of M Street of A County under the condition that the wind direction is SE, the velocity is 2.6 m/s, and the height is 1.5 m. The temperature distribution regularity in the whole district at the height 1.5 m is generally the same, which does not vary much with the solar radiation. The internal temperature of buildings is very high in winter, up to 5 °C. At the meantime, the temperature around buildings is also very high, ranging from 1 to 3 °C. The high temperature generated from underlying surface of buildings is inclined to

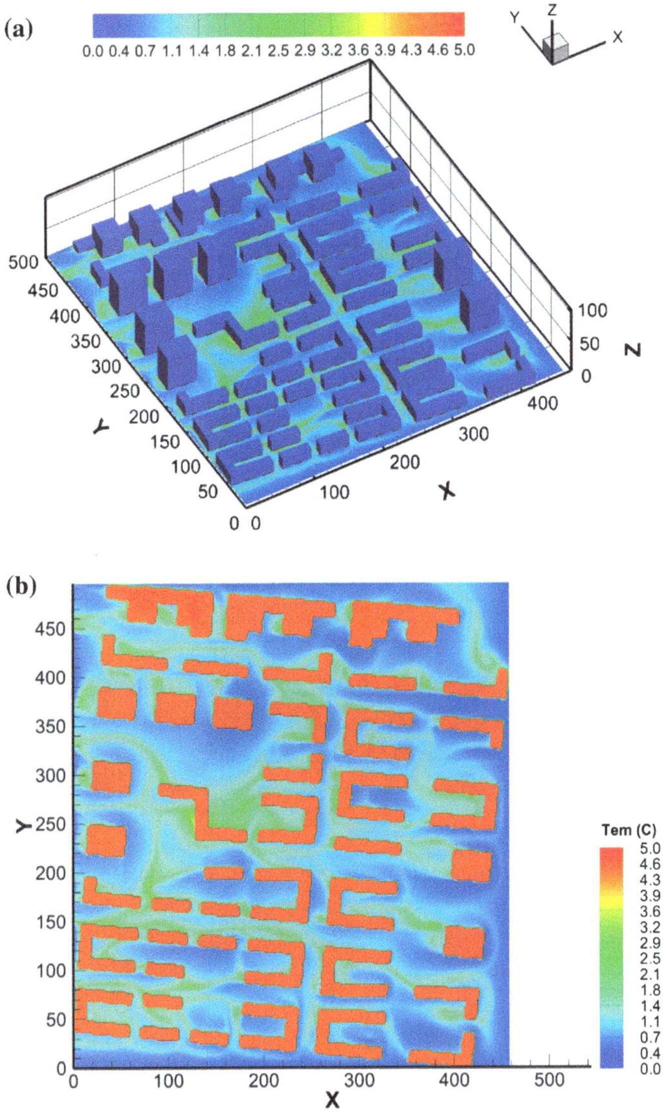

Fig. 3.13 Temperature distribution at the height of 1.5 m

have interactive influence. Furthermore, the inter-space is so narrow that buildings in the district will block the air flow, which leads to the failure of timely heat emission together with air. Therefore, the local temperature among building groups is on a rise. Namely, the building arrangement also has a strong influence on the temperature distribution in the district.

The temperature distribution at the height 1.5 m also has some regularities on the direction of incoming flow, which is specifically reflected in the phenomenon that the

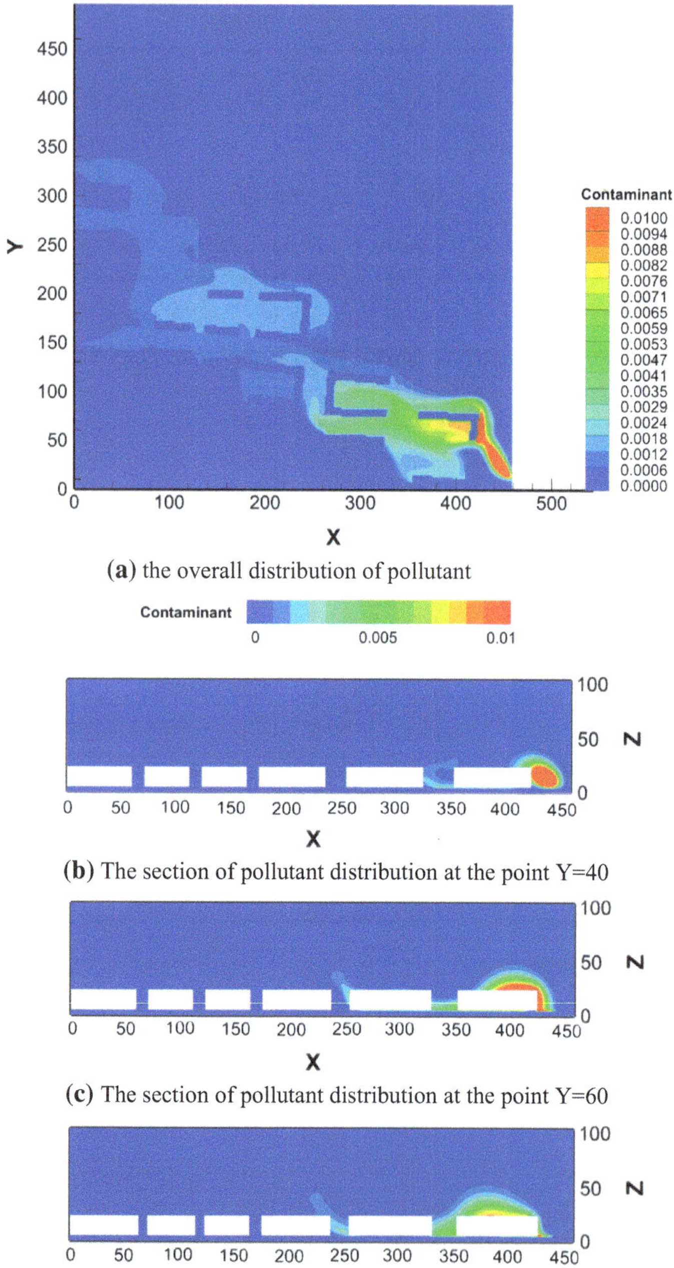

(a) the overall distribution of pollutant

(b) The section of pollutant distribution at the point Y=40

(c) The section of pollutant distribution at the point Y=60

(d) The section of pollutant distribution at the point Y=80

Fig. 3.14 Pollutant distribution at different cross sections local temperature of the incoming wind into the district is much lower than that out of the district. This mainly has much to do with the direction of incoming wind. The heat flows with air and is brought to the downstream on the wind direction. Therefore, the local mean temperature on the wind direction out of the district is much higher.

Figure 3.14 indicates the pollutant distribution of M Street of A County at the points where $Y = 40$, 60, and 80 (the direction of width) under the condition that the wind direction is SE, the velocity is 2.6 m/s, and the height is 1.5 m. It could be seen that on the direction of height (*Z*-axis), the pollutant dispersion, stays within certain range, not exceeding 50 m. In the southeast wind field, the pollutant on the *X*-axis disperses toward west continuously, which leads to a relatively larger dispersion scope, whereas on the *Y*-axis, the pollutant disperses toward north. The concentration decreases with distance. Generally, at the point $Y = 40$, because the wind velocity is insufficient, and the point itself is close to the pollution source, the pollutant climbing is not evident, which leads to the local gathering of pollutant. At the points $Y = 60$ m and $Y = 100$ m, because the concentration is low, and the wind velocity is high, the pollutant dispersion toward downstream is evident. However, the influence of pollutant is limited, which nearly disappears at the distance of 350 m from the pollutant source.

3.4 Conclusions

In this chapter, we conducted a numerical simulation using the commercial software PHOENICS to assess the wind thermal environment and pollutant transport in an urban area. The architectural composition, materials of ground surface, distribution of pollution source, and environmental conditions were considered. The thermal environment and pollutant transport mechanism were analyzed, so were the problems and influential characteristics of current city layout.

References

1. G.-Y. Cao, G. Chen, L.-H. Pang, X.-Y. Zheng, S. Nilsson. Urban growth in China: past, prospect, and its impacts. Population and Environment. 33 (2012) 137–60.
2. H.Q. Cheng, I. Masser. Urban growth pattern modeling: a case study of Wuhan city, PR China. Landscape And Urban Planning. 62 (2003) 199–217.
3. Y. Cai, H. Zhang, P. Zheng, W. Pan. Quantifying the Impact of Land use/Land Cover Changes on the Urban Heat Island: A Case Study of the Natural Wetlands Distribution Area of Fuzhou City, China. Wetlands. 36 (2016) 285–98.
4. H.Y. Du, D.D. Wang, Y.Y. Wang, X.L. Zhao, F. Qin, H. Jiang, et al. Influences of land cover types, meteorological conditions, anthropogenic heat and urban area on surface urban heat island in the Yangtze River Delta Urban Agglomeration. Sci Total Environ. 571 (2016) 461–70.
5. H. Ma, Z. Jiang, J. Song, A. Dai, X. Yang, F. Huo. Effects of urban land-use change in East China on the East Asian summer monsoon based on the CAM5.1 model. Climate Dynamics. 46 (2016) 2977–89.

6. L. Sun, J. Wei, D.H. Duan, Y.M. Guo, D.X. Yang, C. Jia, et al. Impact of Land-Use and Land-Cover Change on urban air quality in representative cities of China. Journal Of Atmospheric And Solar-Terrestrial Physics. 142 (2016) 43–54.
7. R. Yu, Z. Jiang, P. Zhai. Impact of urban land-use change in eastern China on the East Asian subtropical monsoon: A numerical study. Journal Of Meteorological Research. 30 (2016) 203–16.
8. L.F. Han, Y.P. Xu, G.B. Pan, X.J. Deng, C.S. Hu, H.L. Xu, et al. Changing properties of precipitation extremes in the urban areas, Yangtze River Delta, China, during 1957–2013. Nat Hazards. 79 (2015) 437–54.
9. H. Zhang, X.M. Jing, J.Y. Chen, J.J. Li, B. Schwegler. Characterizing Urban Fabric Properties and Their Thermal Effect Using QuickBird Image and Landsat 8 Thermal Infrared (TIR) Data: The Case of Downtown Shanghai, China. Remote Sens. 8 (2016) 20.
10. W. Li, Y. Bai, W. Zhou, C. Han, L. Han. Land Use Significantly Affects the Distribution of Urban Green Space: Case Study of Shanghai, China. Journal Of Urban Planning And Development. 141 (2015).
11. J. Qian, Y. Peng, C. Luo, C. Wu, Q. Du. Urban Land Expansion and Sustainable Land Use Policy in Shenzhen: A Case Study of China's Rapid Urbanization. Sustainability. 8 (2016).
12. P. Starke, P. Goebel, W.G. Coldewey. Urban evaporation rates for water-permeable pavements. Water Science And Technology. 62 (2010) 1161–9.
13. X. Chen, N. Zhang, F. Wu. Impacts of Rainfall and Land Use on Urban Surface Runoff: A Case Study of Area Surrounding the North Moat in Beijing, China. Journal of Natural Resources. 29 (2014) 1391–402.
14. C.S.B. Grimmond, T.R. Oke. Heat storage in urban areas: Local-scale observations and evaluation of a simple model. Journal Of Applied Meteorology. 38 (1999) 922–40.
15. V. John, P. Jain, M. Rahate, P. Labhasetwar. Assessment of deterioration in water quality from source to household storage in semi-urban settings of developing countries. Environmental Monitoring And Assessment. 186 (2014) 725–34.
16. C.H. Liao, C.L. Chang, C.Y. Su, P.T. Chiueh. Correlation between land-use change and greenhouse gas emissions in urban areas. International Journal Of Environmental Science And Technology. 10 (2013) 1275–86.
17. N. Kammuang-Lue, P. Sakulchangsatjatai, P. Sangnum, P. Terdtoon. INFLUENCES OF POPULATION, BUILDING, AND TRAFFIC DENSITIES ON URBAN HEAT ISLAND INTENSITY IN CHIANG MM CITY, THAILAND. Thermal Science. 19 (2015) S445–S55.
18. R. Kotharkar, M. Surawar. Land Use, Land Cover, and Population Density Impact on the Formation of Canopy Urban Heat Islands through Traverse Survey in the Nagpur Urban Area, India. Journal Of Urban Planning And Development. 142 (2016).
19. J. Qian, H. Liu, H. Lin, l. Zhu, X. Wang, W. Jiang. Numerical simulation of the impact of urban growth on Suzhous urban haze. Journal of Nanjing University Natural Sciences. 51 (2015) 551–61.
20. C. Cao, X. Lee, S. Liu, N. Schultz, W. Xiao, M. Zhang, et al. Urban heat islands in China enhanced by haze pollution. Nature Communications. 7 (2016).
21. Z.R. Liu, Y.S. Wang, B. Hu, D.S. Ji, J.K. Zhang, F.K. Wu, et al. Source appointment of fine particle number and volume concentration during severe haze pollution in Beijing in January 2013. Environ Sci Pollut Res. 23 (2016) 6845–60.
22. J. Liu, J. Li, M. Vonwiller, D. Liu, H. Cheng, K. Shen, et al. The importance of non-fossil sources in carbonaceous aerosols in a megacity of central China during the 2013 winter haze episode: A source apportionment constrained by radiocarbon and organic tracers. Atmospheric Environment. 144 (2016) 60–8.

Chapter 4
Filed Synergy Analysis on Pollutant Dispersion in Built Environment

4.1 Introduction

Guo et al. [1] firstly put forward the concept of field synergy between the velocity distribution and the temperature distribution. They found that during the convective heat transfer, the field synergy capacity is relevant to the fluid heat transfer, based upon which, they proposed the Field Synergy Principles (FSP) which provided a new vista for the performance assessment of heat transfer enhancement. The FSP of convective heat transfer was verified via a series of experiments data [2–6]. It is different from the conventional convective heat transfer enhancement in which both the flow resistance and the pumping power are on an obvious increase. However, the convective heat transfer enhancement under the FSP could reduce the amplification of pumping power during the enhancement, which is conductive to the engineering application. The field synergy principle is widely applied in engineering. Lots of new techniques such as the slotted fin for enhancing heat transfer [7], sharp expansion, and accelerated flow converging-diverging tube [8] have come into being. Spirally Yan-yang et al. [9] managed to optimize the pulse tube refrigerator [10] and shell-and-tube heat exchanger [11] based upon the new techniques derived from FSP.

Recently, novel methods of heat transfer enhancement and optimization were reported to consider the effect of enhanced heat transfer tubes. Main considerations were focused on the center of the tubes by inserting porous medium with high conductivity, high porosity, and high filling ratio [12–16], various compound baffles [17–22], twist tapes [14, 17–19, 22–28], and discrete double-inclined ribs [27–30]. Recently, the minimum power consumption and minimum entransy dissipation principles were proposed to assess the effect of enhancing convective heat transfer [31, 32], and the active optimization design theory based on the exergy destruction minimization, local exergy destruction rate, and local exergy destruction

T. Ming et al., *Pollutant Dispersion in Built Environment*,
DOI 10.1007/978-981-10-3821-1_4

minimization was advanced to get heat transfer tubes with excellent performances [33–42]. The active optimization design theory has been extended to various applications from refrigerators and heat engines [39, 43–48] to waste heat recovery systems [40–42, 49].

Based on the principle of field synergy for heat transfer enhancement [1], Liu et al. [50] and Wei et al. [51] proposed the concept of physical quantity synergy in the laminar flow field according to the physical mechanism of convective heat transfer between fluid and tube wall. The synergy regulation among physical quantities of fluid particle was revealed by establishing formulas reflecting the relation between synergy angles and heat transfer enhancement. The physical nature of enhancing heat transfer and reducing flow resistance, which is directly associated with synergy angles α, β, γ, Φ, θ, and Ψ was also explained. The principle of synergy among physical quantities was numerically verified by the calculation of heat transfer and flow in a thin cylinder-interpolated tube. Later, the concept of physical quantity synergy was extended to turbulent flow which was named as multi-field synergy principle [52] and verified by various cases of enhanced heat transfer tubes [17, 18, 21, 24, 25, 27, 28, 30, 53–59].

Due to the similarity between Fourier's Law of Conduction and Fick's Law of Diffusion, and that between mass transfer and heat transfer, Chen et al. [60] extended FSP to mass transfer. They advocated that the synergy angle of velocity distribution with concentration distribution may influence the concentration of pollutant. They defined mass entransy dissipation and verified the convective mass transfer synergy principles via the actual mass transfer inside the two-dimension rectangular cell. Researches relevant to the application of Mass Transfer Synergy Principles (MTSP) are often seen in the literatures, such as improving the performance of dehumidifying air conditioner [61], guiding the structural design of catalytic reactor [62], and optimizing the ventilation and drainage inside the space station [11], which has provided an interpretation for the mechanism of spiral tube convective mass transfer enhancement [63].

The essence of pollutant dispersion in fluid flow is a kind of convective mass transfer. Based on the analogy between heat transfer and mass transfer and with the reference to the field synergy analysis on the convective heat transfer during the mass transfer of pollutant dispersion, it could be seen that the synergy angle of velocity with temperature gradient determines the overall performance of heat transfer, which could further be the basis for the determination of heat transfer entransy dissipation. In this chapter, we extended the heat transfer field synergy principle to mass transfer field synergy. With the comparison between temperature gradient and concentration gradient, we managed to analyze the synergy angle of velocity with concentration gradient, based upon which, we also revealed the relationship between synergy angle and pollutant concentration. Further, via the mass entransy dissipation derived from the heat transfer function, we analyzed the relationship between that and pollutant concentration. Under various boundary conditions (inlet velocity and pollutant concentration of the source) of different structures, we analyzed the synergy angle between velocity and concentration gradient on the pollutant and explore the influence parameters.

4.2 Field Synergy Principle of Mass Transfer

Based upon the component concentration equation of convective mass transfer without source component item at the static state, we integrated the equation via Gauss theorem, ignoring the longitudinal diffusion of inlet and outlet components, and we got:

$$\int_{\Omega} \rho |U||\nabla C| \cos\beta \, \mathrm{d}V = \int_{S} \vec{n} \cdot (\rho D \nabla C) \mathrm{d}S, \tag{4.1}$$

where ρ is the density (kg/m^3), U is the fluid velocity (m/s), ∇C stands for the component concentration gradient, D is the mass diffusion coefficient (m^2/s), and β is the included angle which is called as the synergy angle formed by the velocity vector and the component concentration. The smaller the included angle is, the better the synergy is, the larger the mass transfer of the system with the surroundings is, and the lower the pollutant concentration inside the system is.

Chen et al. [60] proposed the convective mass transfer field synergy numbers and defined the equivalent characteristic length as:

$$L = \frac{V}{S}. \tag{4.2}$$

Meanwhile, the nondimensional velocity, pollutant concentration gradient, and infinitesimal volume can be defined as:

$$\overline{U} = \frac{U}{U_{\mathrm{in}}}, \tag{4.3}$$

$$\nabla\overline{C} = \frac{\nabla C}{\frac{C_{\mathrm{ms}} - C_{\mathrm{in}}}{L}}, \tag{4.4}$$

$$\mathrm{d}\overline{V} = \frac{\mathrm{d}V}{V}, \tag{4.5}$$

where V and S are the volume and superficial area of the convective mass transfer domain.

When we substitute the above nondimensional variables into Eq. (4.5),

$$\int_{\Omega} \rho \vec{U} \cdot \nabla C \, \mathrm{d}V = \int_{S} \vec{n} \cdot (\rho D \nabla C) \mathrm{d}S. \tag{4.6}$$

We could get the nondimensional Sherwood number for the convective mass transfer:

$$Sh = Re\, Sc \int_{\Omega} \overline{U} \cdot \nabla \overline{C} \, \mathrm{d}\overline{V} \tag{4.7}$$

where $Sh = \frac{h_{\mathrm{m}} d}{D}$ is the Sherwood number, $Sc = \frac{\nu}{D}$ is the Schmidt number which correlates with the fluid property, and h_{m} is the coefficient of convective mass transfer.

Seen from Eq. (4.7), the Sh number correlates with not only Reynolds number and Schmidt number but also the synergy angle of velocity vector and concentration gradient vector, namely the value of integration of $\overline{U} \cdot \nabla \overline{C}$ in the fluid domain.

So, the field synergy number of convective mass transfer Fc_{m} can be defined as:

$$Fc_{\mathrm{m}} = \int_{\Omega} \overline{U} \cdot \nabla \overline{C} \, \mathrm{d}\overline{V}. \tag{4.8}$$

Meanwhile, to illustrate the energy loss during the mass transfer, the concept of mass entransy dissipation was introduced, which could be defined as below:

$$\psi = 0.5\, \rho D \nabla C \cdot \nabla C \tag{4.9}$$

If we integrated the mass entransy dissipation in the fluid domain, we can obtain the dissipative function:

$$Z = \int_{\Omega} 0.5 \rho D \nabla C \cdot \nabla C \, \mathrm{d}\Omega \tag{4.10}$$

From Eq. (4.8), $\overline{U}$ and $\nabla \overline{C}$ indicate nondimensional velocity vector and nondimensional concentration gradient, respectively. The dot product of nondimensional velocity and concentration gradient can be written as:

$$\overline{U} \cdot \nabla \overline{C} = \left|\overline{U}\right| \left|\nabla \overline{C}\right| \cos\theta. \tag{4.11}$$

Obviously, the smaller the included angle θ between ∇C and U, the larger the dot product $\overline{U} \cdot \nabla \overline{C}$ is, and the larger the field synergy number Fc_{m} is.

The mass transfer field synergy angle θ could be:

$$\theta = \arccos\left(\frac{\overline{U}\cdot\nabla\overline{C}}{|\overline{U}||\nabla\overline{C}|}\right). \tag{4.12}$$

4.3 Physical and Mathematical Model

4.3.1 Physical Model

The, geometric models, and boundary conditions in computation are shown schematically in Fig. 4.1. For the two-dimensional model, five street canyons are formed by six rows of buildings with the same height *H*, as shown in Fig. 4.1a; the

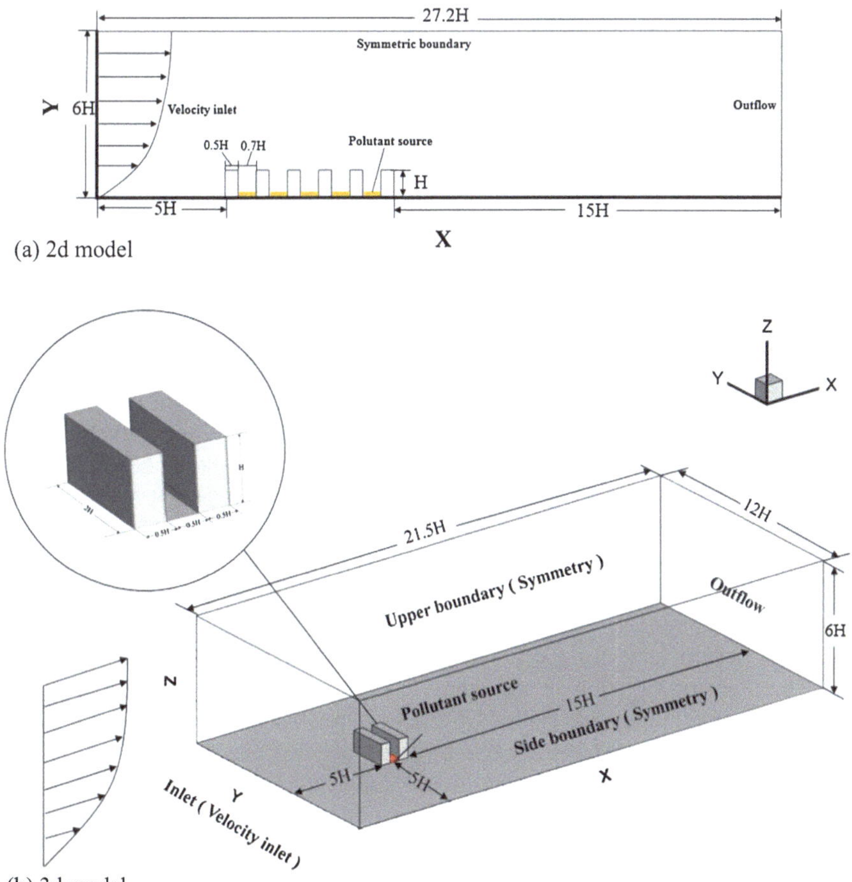

Fig. 4.1 Computational domain and boundary conditions: **a** 2D model; **b** 3D model

distance between the ambient crosswind inlet and the leading edge of the first building is set as 5*H*. The distance between outflow boundary and leeward edge of the last building is 15.5*H*. The upper boundary is set as 6*H* to the ground surface. For three-dimensional model shown in Fig. 4.1b, the domain was extended up to a distance of 5*H* upwind of the first building (*H*: building height), 15.5*H* downwind of the last building, 5*H* in each crosswind direction, and 5*H* in the vertical direction, according to the AIJ guidelines for practical applications of CFD.

4.3.2 Mathematical Model

As mentioned, the flow field, if the perpendicular approaching wind is relative small, is considered to be purely natural convection induced by solar radiation which heats the ground surface and building walls and then results in the increase of air temperature. Richardson number is used to represent the importance of natural convection relative to the forced convection. However, the approaching wind plays a significant role in previous investigations as the buildings are all exposed to the ambient, which causes the system to be a combination of natural and forced convection. The natural flow should be vertically upward due to the effect of horizontal ground surface having high temperature; however, the airflow horizontally enhances this heat transfer effect, which puts the ground surface at a lower temperature level. The flow regimes can also be classified according to the ratio of *Gr* and Re^2 shown as:

$$\frac{Gr}{\mathrm{Re}^2} = \frac{(g\beta\Delta TL^3/v^2)}{(U_{\mathrm{ref}}L/v)^2} = \frac{g\beta\Delta TL}{U_{\mathrm{ref}}^2}. \tag{4.13}$$

where, *Gr* and *Re* are Grashof and Reynolds numbers, respectively; ΔT is the maximum temperature increase within the area; v, β, and L are, respectively, the air viscosity, thermal expansion coefficient, and the maximum building height; g is the acceleration of gravity. U_{ref} is the wind velocity at reference height in the area.

After the preliminary calculation, we found that *Gr* is higher than 10^{10} which is the upper limit transition value of natural convection from laminar to turbulent flow. Therefore, the turbulent mathematical model needs to be selected to describe fluid flow within the system. The RNG $k - \varepsilon$ turbulence model is chosen to evaluate numerical performance for the pollutant dispersion in an isolated street canyon. Also, the density variation in the whole computational model is so small. The density variation of the air is caused by temperature changes, rather than that of the pressure. The incompressible flow is assumed and we use the ideal gas law to express the relationship between density and temperature for natural convection. As a result, the transport equations for incompressible turbulent flow can be written as follows:

Mass equation:

$$\frac{\partial \rho}{\partial t} + \frac{\partial}{\partial x_i}(\rho u_i) = 0. \tag{4.14}$$

Navier–Stokes equation:

$$\frac{\partial}{\partial t}(\rho u_i) + \frac{\partial}{\partial x_j}(\rho u_i u_j) = -\frac{\partial p}{\partial x_i} + \frac{\partial \tau_{ij}}{\partial x_j} + \rho g_i + F_i, \tag{4.15}$$

where the stress tensor τ_{ij} is defined as:

$$\tau_{ij} = \left[\mu\left(\frac{\partial u_i}{\partial x_j} + \frac{\partial u_j}{\partial x_i}\right)\right] - \frac{2}{3}\mu\frac{\partial u_i}{\partial x_i}\delta_{ij} \tag{4.16}$$

Energy equation:

$$\frac{\partial}{\partial t}(\rho E) + \frac{\partial}{\partial x_i}(u_i(\rho E + p)) = \frac{\partial}{\partial x_j}\left(k_{\text{eff}}\frac{\partial T}{\partial x_j} - \sum_j h_j J_j + u_j(\tau_{ij})_{\text{eff}}\right) + S_{\text{h}} \tag{4.17}$$

where k_{eff} is the effective conductivity (k + k_{t}, where k_{t} is the turbulent thermal conductivity); J_j is the diffusion flux of species j; S_{h} includes the heat of chemical reaction or any other volumetric heat sources. $(\tau_{ij})_{\text{eff}}$ is the deviatoric stress tensor, defined as:

$$(\tau_{ij})_{\text{eff}} = \mu_{\text{eff}}\left(\frac{\partial u_j}{\partial x_i} + \frac{\partial u_i}{\partial x_j}\right) - \frac{2}{3}\mu_{\text{eff}}\frac{\partial u_k}{\partial x_k}\delta_{ij} \tag{4.18}$$

In Eq. (4.8):

$$E = h - \frac{p}{\rho} + \frac{v^2}{2} \tag{4.19}$$

where sensible enthalpy h is defined for incompressible flows as:

$$h = \sum_j Y_j h_j + \frac{p}{\rho} \tag{4.20}$$

where Y_j is the mass fraction of species j.

Equation for the turbulent kinetic energy k:

$$\frac{\partial}{\partial t}(\rho k) + \frac{\partial}{\partial x_i}(\rho k u_i) = \frac{\partial}{\partial x_j}\left[\left(\mu + \frac{\mu_t}{\sigma_k}\right)\frac{\partial k}{\partial x_j}\right] + G_k + G_{\text{b}} - \rho\varepsilon - Y_M + S_k \tag{4.21}$$

Equation for the energy dissipation:

$$\frac{\partial}{\partial t}(\rho\varepsilon) + \frac{\partial}{\partial x_i}(\rho\varepsilon u_i) = \frac{\partial}{\partial x_j}\left[\left(\mu + \frac{\mu_t}{\sigma_\varepsilon}\right)\frac{\partial \varepsilon}{\partial x_j}\right] + C_{1\varepsilon}\frac{\varepsilon}{k}(G_k + C_{3\varepsilon}G_b) - C_{2\varepsilon}\rho\frac{\varepsilon^2}{k} + S_\varepsilon \tag{4.22}$$

where G_k represents the generation of turbulence kinetic energy because of the mean velocity gradients and can be defined as $G_k = -\rho\overline{u_i' u_j'}\left(\partial u_j/\partial x_i\right)$; G_b is the generation of turbulence kinetic energy due to buoyancy; σ_k and σ_ε are the turbulent Prandtl numbers for k and ε: $\sigma_k = 1.0$, $\sigma_\varepsilon = 1.3$. $C_{1\varepsilon}$, $C_{2\varepsilon}$ are constants: $C_{1\varepsilon} = 1.44$, $C_{2\varepsilon} = 1.92$. $\mu_t = \left(C_\mu \rho k^2/\varepsilon\right)$ and $C_\mu = 0.09$.

The species (pollutants) transport equation:

$$\frac{\partial C_i}{\partial t} + u_j\frac{\partial C_i}{\partial x_j} = \frac{1}{\rho}\frac{\partial}{\partial x_j}\left[\left(D_i + \frac{\mu_t}{Sc_t}\right)\frac{\partial C_i}{\partial x_j}\right] \tag{4.23}$$

where C_i is the species (pollutants) concentration; D_i is the diffusivity; Sc_t is the turbulent Schmidt number.

The RNG approach is a mathematical technique that can be used to derive a turbulence model similar to the k-epsilon; it has a modified form of the epsilon equation which attempts to account for the different scales of motion through changes to the production term. In addition, the RNG k–ε turbulence model has analytically derived differential formulas for determining the effective viscosity and the turbulent Prandtl number instead of a constant value for the standard k–ε turbulence model as follows:

$$\frac{\partial}{\partial t}(\rho k) + \frac{\partial}{\partial x_i}(\rho k u_i) = \frac{\partial}{\partial x_j}\left[\alpha_k \mu_{\text{eff}}\frac{\partial k}{\partial x_j}\right] + G_k + G_b - \rho\varepsilon - Y_M + S_k \tag{4.24}$$

$$\begin{aligned}\frac{\partial}{\partial t}(\rho\varepsilon) + \frac{\partial}{\partial x_i}(\rho\varepsilon u_i) = {} & \frac{\partial}{\partial x_j}\left[\alpha_\varepsilon \mu_{\text{eff}}\frac{\partial \varepsilon}{\partial x_j}\right] + C_{1\varepsilon}\frac{\varepsilon}{k}(G_k + C_{3\varepsilon}G_b) \\ & - \left[\rho C_{2\varepsilon} + \frac{C_\mu \rho \eta^3(1 - \eta/\eta_0)}{1 + \beta\eta^3}\right]\frac{\varepsilon^2}{k} + S_\varepsilon\end{aligned} \tag{4.25}$$

where α_k and α_ε are the inverse effective Prandtl number for k and ε, respectively; μ_{eff} is the effective turbulent viscosity; $\eta = Sk/\varepsilon$, S is the scalar measure of the deformation tensor; the constants of η_0 and β are 4.38 and 0.012, respectively.

4.3.3 Calculation of Solar Radiation

For two-dimensional model, the temperature at land surface is assumed constant and higher than that of incoming wind flow, which could be valid due to strong solar radiation and low heat capacity of urban road materials. For three-dimensional model, solar radiation will be imposed on the ground surface and all the building surfaces, varying with time, solar radiation (I_c) on an arbitrary tilted surface having a tilt angle of β from the horizontal and an azimuth angle of α_w is the sum of components consisting of beam ($I_{b,c}$), sky-diffused ($I_{d,c}$), and ground-reflected solar radiation ($I_{r,c}$):

$$I_c = I_{b,c} + I_{d,c} + I_{r,c} \tag{4.26}$$

$$I_{b,c} = I_{b,N} \cos i = I_{b,N}[\cos \alpha \cos(\alpha_s - \alpha_w) \sin \beta + \sin \alpha \cos \beta] \tag{4.27}$$

$$I_{d,c} = CI_{b,N} \cos^2\left(\frac{\beta}{2}\right) \tag{4.28}$$

$$I_{r,c} = \gamma I_{b,N}(\sin \alpha + C) \sin^2\left(\frac{\beta}{2}\right) \tag{4.29}$$

$$I_{b,N} = C_n I e^{-L/\sin \alpha} \tag{4.30}$$

where C, C_n are constants related to the sky conditions, L is the optical depth for average atmospheric condition at sea level with a moderately dusty atmosphere, and γ is the surface reflectance.

In Eq. (4.30), the solar radiation I can be calculated as:

$$I = I_0\left(\frac{D_0}{D}\right)^2 \tag{4.31}$$

where I_0 is the solar constant, D is the distance between the sun and the earth, and D_0 is the yearly mean earth–sun distance.

4.3.4 Boundary Conditions

(1) **Inlet boundary**

For 3D model, $U(x)$ and $U(y)$ are considered to be equal to zero. And the wind velocity profile in both 2D and 3D model is provided by:

$$\mathrm{U}(z) = U_{\mathrm{ref}}\left(\frac{z}{0.02}\right)^{0.2} \tag{4.32}$$

Turbulent kinetic energy and dissipation rate at inlet boundary were specified as:

$$k = \frac{U_*^2}{\sqrt{C_\mu}}, \tag{4.33}$$

$$\varepsilon = \frac{U_*^3}{\kappa z}, \tag{4.34}$$

where U_{ref} represents the velocity at a reference height. U_* represents the friction velocity. κ is the von Karman constant (0.4), and $C_\mu = 0.09$.

(2) **Outlet boundary**

Outflow boundary condition is applied to each of these and the simplification that reversed flow is normal to the boundary surface. This boundary condition combines a zero diffusion flux for all variables with an overall mass balance correction, which is governed by the following equation:

$$\partial/\partial x(u, v, w, k, \varepsilon) = \partial/\partial y(u, v, w, k, \varepsilon) = \partial/\partial z(u, v, w, k, \varepsilon) = 0, \tag{4.35}$$

where x is the coordination of flow direction.

(3) **Wall boundary**

For 3D model, both the ground surface and all the surfaces of the buildings will receive solar radiation, and radiation heat transfer will occur between any two surfaces.

In addition, the ground and building surface will exchange radiation heat with the sky and affected by the shadow angles and view factors. Solar radiation is imposed on the ground and all the building surfaces, varying with time, shape angle, view factor, and solar radiation intensity. Whereas in 3D model, the solar radiation model is adopted, the heat transfer coefficient between the building surfaces and air is assumed as 25 W/(m^2 K).

For 2D model, the temperature at land surface is assumed constant and higher than that of incoming wind flow, which could be valid due to strong solar radiation and low heat capacity of urban road materials.

Enhanced wall treatment and zero roughness are used on all solid boundaries, i.e., building roofs, facades, and streets. The wall functions in CFD coding are based on the universal law of the wall. In ANSYS Fluent, the wall function (law of the wall for mean velocity) for smooth surfaces is written as:

$$\frac{U_p u^*}{\tau_0/\rho} = \frac{1}{\kappa}\ln(Ey^*) \tag{4.36}$$

where the nondimensional distance y^* is calculated by $y^* = \rho u^* y_p/\mu_t$, which is approximately equal to y^+ $\left(y^+ = \rho u y_p/\mu_t\right)$ in the equilibrium of turbulent boundary layers.

(4) **Other boundary**

At the upper boundary and at the lateral boundary (3D), symmetry boundary condition is applied, i.e., zero normal gradient. The pollutant release at pedestrian level (1–2 m) is represented by a source zone Q at the street canyon.

4.3.5 *Meshing Skills*

In general, for the same meshing zone, the hexahedral (HEX) meshing method is more economical and can reduce false diffusion more efficiently than the tetrahedral method. As a result, HEX grids were applied in the model and the mesh generation procedure of the whole geometric model was executed using the commercial software package Gambit 2.4.6. Figure 4.2 reveals the grid distribution of the geometric models. Because of the anticipated relatively steep gradients in velocity, pressure, and temperature near the walls and the ground, serious consideration should be given to the boundary layer effect of these areas. Thereby, the fine grids near the building walls and ground surfaces were more concentrated in these locations.

4.3.6 *The Numerical and Computational Model*

To verify the relationship between the field synergy and the intensity of convective mass transfer during the pollutant dispersion, We firstly established five two-dimension planar models, with the width of different streets the variable in the paper, Four cases were designed for each *H*/*D* ratio (building height vs. street canyon width), with a total of 16 numerical tests carried out to investigate the influence of thermal effects with different ambient wind velocities. The research objects consist of six rows of buildings and five street canyons. For each *H*/*D* ratio, the ambient crosswind velocities at the height of 10 m in the four numerical tests

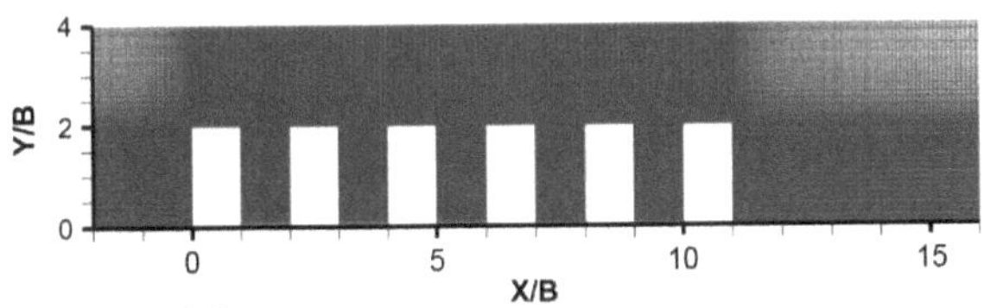

(c) 2D model

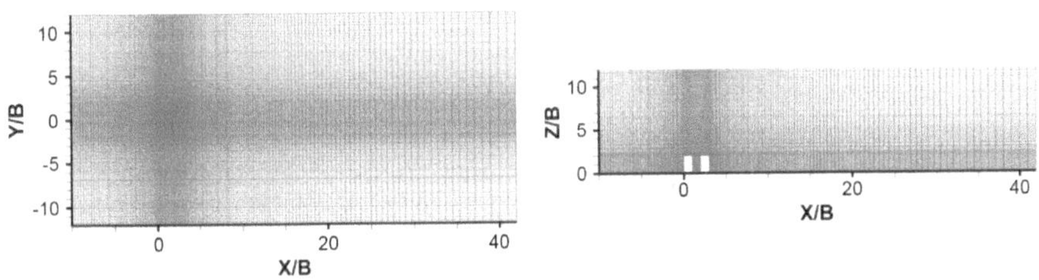

(d) 3D model

Fig. 4.2 Grid distribution of the geometric models

are 2.0, 3.0, 4.0, and 5.0 m/s, respectively, and the ΔT is set as 20 K for each test. According to the guidelines of COST Action 732 [64], the upstream distance from the inlet plane to the building is set to $5H$, and the downstream distance from the outflow plane to the building is set to $15H$.

4.4 Results and Discussions

4.4.1 2D Model Analysis

In order to verify the mass transfer field synergy principle, investigations were conducted to consider street canyons under thermal wall conditions and two different building height: H = 10 m; H = 14 m. These applications are dedicated to observing the calculated results of convective mass transfer mass and mass transfer field synergy number at different ambient crosswind velocities, and the relationship between convective mass transfer performance and the effect of convective mass transfer field synergy will be explored further. For each building height H, the ambient crosswind velocities at the height of 10 m above the building in the four numerical cases are 2.0, 3.0, 4.0, 5.0 m/s, respectively. To simplify the 3D model to a 2D one, the ACV direction was given normal to street canyon. The inlet is on the left side of the domain, and the measured velocity profile in the upstream free flow is used to provide the inlet boundary condition which can be approximately described by a power-law profile. The highly mixing homogeneity of pollutants released by vehicles and air is due to the motion of pedestrian and vehicles, and the perfectly mixed near-ground pollutant emission is described as uniform volume pollutant sources. Based on this assumption, the initial region of the emissions is at the pedestrian level from z = 0 m to z = 1 m at the bottom of the street canyon. To reduce disturbance generated by pollutant release, the intense of pollutant source at the pedestrian level is set to 5e−7 kg/m^3 s as the minimum. The ambient inlet concentration is zero. The diffusion coefficient of the gaseous pollutant D in the air is 2.88e−5m^2/s. In order to define mass transfer field synergy number, the height of building H is given as the characteristic length; the ambient crosswind velocity (ACV) characteristic velocity is given as the characteristic velocity; the mass concentration difference between concentration of inlet and pollutant emissions is given as the characteristic mass concentration difference.

4.4.1.1 Characteristics of the Flow Field for Various Street Widths in 2D Model

Figure 4.3 shows the normalized concentration, streamline, velocity, and mass transfer field synergy angle distributions in the first street at D = 10 m. When the ambient crosswind passes by the first street, two vortexes of opposite directions are

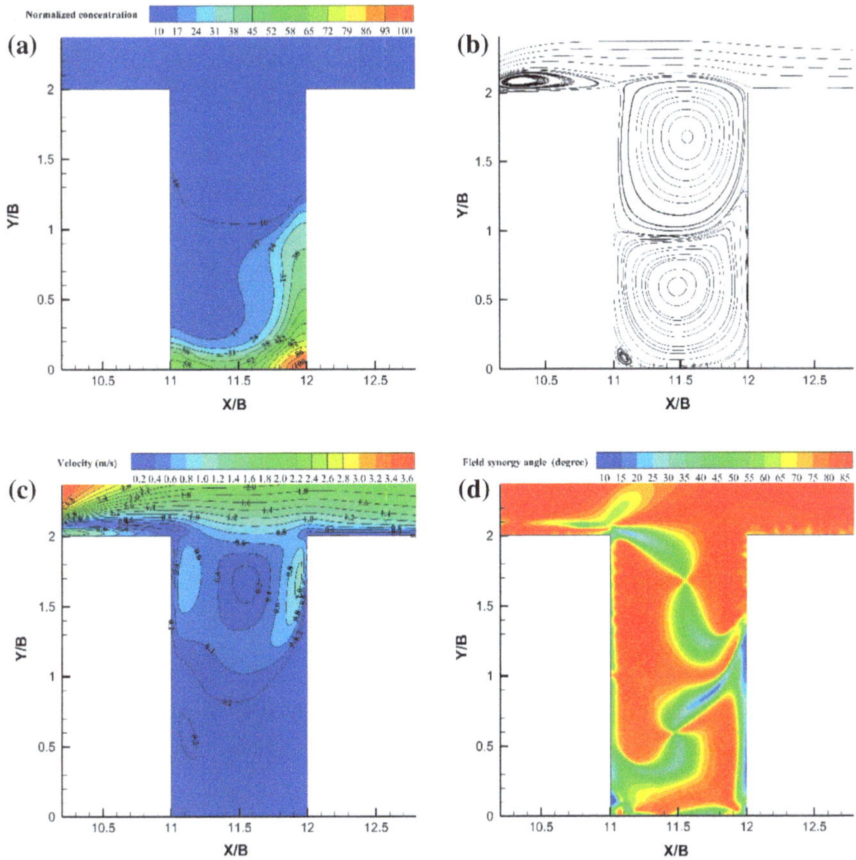

Fig. 4.3 Distributions in first street canyon (U_{ref} = 2 m/s, D = 10 m): **a** normalized concentration; **b** streamline; **c** velocity; **d** mass transfer field synergy angle

generated inside the street canyon. Seen from the bottom, upper left, and low right of the canyon in Fig. 4.3a, b, the trend of fluid velocity nearly parallel with iso-concentration fraction bar, which indicates that the velocity vector is perpendicular to the concentration gradient. From Fig. 4.3d, we can find that the synergy angles at the bottom, upper left, and lower right all fall in between 75° and 85°. Thereby, the synergy of velocity distribution and concentration distribution is very poor. At the meantime, for each vortex, there exist two regions whose synergy are good, with two central symmetries and another two with poor synergy. It could be seen clearly from Fig. 4.3a that in the vortex upside, the concentration gradient is the perpendicularity of the left lower quarter of the vortex down to the center. Therefore, the trend of fluid velocity at the left lower quarter and at the right upper quarter parallel to the iso-concentration fraction bar, and the velocity vector is perpendicular to the concentration gradient. Thus the synergy is relatively better. Similarly, in the vortex downside, since the concentration gradient is the

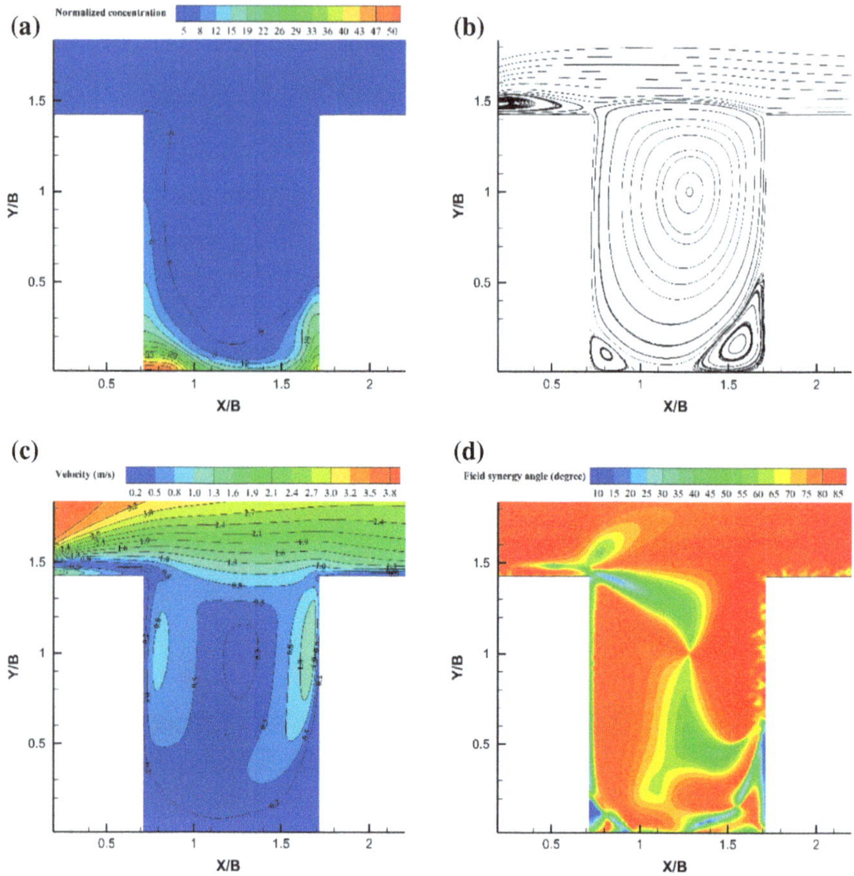

Fig. 4.4 Distribution in first street canyon (U_{ref} = 2 m/s, D = 14 m): **a** normalized concentration; **b** streamline; **c** velocity; **d** mass transfer field synergy angle

perpendicularity of the lower right quarter down to the center, the synergy at the lower right quarter and at the upper left quarter is poor whereas that at the lower left and at the upper right quarter is good. As a result, according to FSP, the flow velocity and the concentration gradient in the street canyon would influence the degree of synergy.

Figure 4.4 intercepted the first street at D = 14 m as the object of analysis. Comparing Fig. 4.4a, b, we could find that the extremum normalized concentration of pollutant at the street section D = 10 m is 11.4, whereas that at D = 14 m is 8.4. Therefore, the pollutant concentration at D = 14 m has been reduced drastically. According to FSP, in comparison with the velocity at D = 10 m and at D = 14 m, it is clear that the velocity at D = 14 m is higher, which also with larger concentration gradient, the synergy effect of velocity distribution, and concentration distribution is better, thus makes the pollutant concentration at D = 14 m much lower.

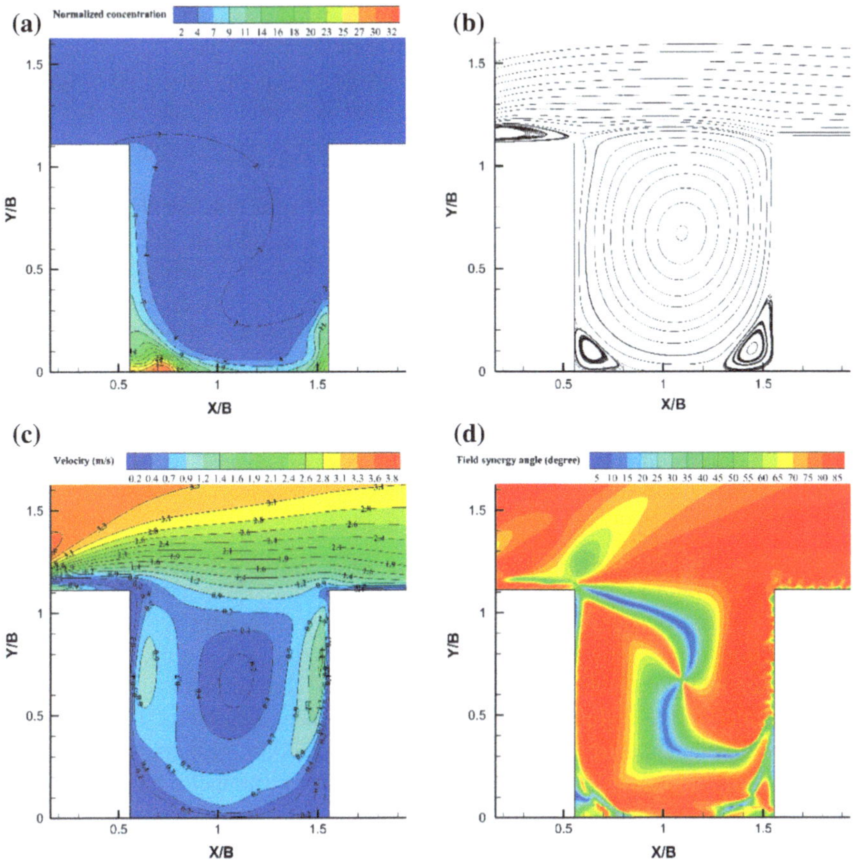

Fig. 4.5 Distribution in first street canyon (U_{ref} = 2 m/s, D = 18 m): **a** normalized concentration; **b** streamline; **c** velocity; **d** mass transfer field synergy angle

Figures 4.3, 4.4, and 4.5 indicate the inlet characteristic velocity distribution of the first street whose widths are D = 10 m, D = 14 m, and D = 18 m, respectively. From the three figures we could find that when the street width increases, the normalized concentration extremum in the flow field decreases gradually, indicating that the increase of pollutant mass transfer, which further verifies the results in Fig. 4.4. Via the comparison of the velocity distributions among those three figures, we could find that the larger the street width is, the much more easily the ambient crosswind will penetrate into the canyon, which disturbs the flow field. In the street canyon D = 10 m, the fluid velocity below the 10 m level is less than 0.2 m/s whereas in the canyon D = 14 m and D = 18 m, the area of the flow field whose velocity is below 0.2 m/s further shrinks. Combined the definition of mass transfer synergy number and FSP, the larger the velocity vector is, the larger the mass transfer synergy number is. The convective mass transfer is intensified, and it will lead to the decrease of the mean concentration in the flow field.

Meanwhile, seen from the distribution of field synergy angle, the synergy between velocity vector and concentration gradient near buildings and the street surface at both sides of the street canyon is poor, approximate to perpendicularity. It could be observed that the flow field velocity near the windward side of street canyon is obviously higher than that near the leeward side, whereas the distribution of the synergy angle of mass transfer field near the windward side is obviously smaller than that near the leeward side. It indicates that the synergy level of mass transfer field at the windward side in the canyon is higher than that at the leeward side. This makes the accumulation of pollutant much easier at the leeward side, which confronts the residents' health there with greater threats.

4.4.1.2 Analysis of the Mass Field Synergy for Various ACV in 2D Model

Figure 4.6 is the variation of pollutant mass fraction with ACV. With increasing ACV, the pollutant mass fraction increases nonlinearly. From the perspective of convection mass transfer, the increase of the ambient crosswind velocity results in the reduction of the flow stability and the convective mass transfer coefficient between air and pollutant. The decreased pollutant concentration means that the dilution effect on the pollutant (CO) increases obviously, but the dilution effect is a qualitative explanation from the perspective of convection mass transfer.

Figures 4.7 and 4.8 show the variation of mass transfer field synergy number and mass entransy dissipation with ACV. Further quantitative calculations should be made as well to assess the intensity of dilution effect. Thus, the mass entransy dissipation function and mass transfer field synergy numbers are introduced to

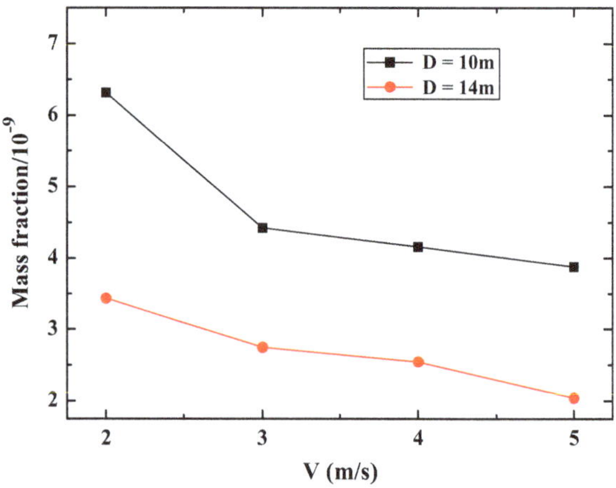

Fig. 4.6 Variation of pollutant mass fraction with ACV

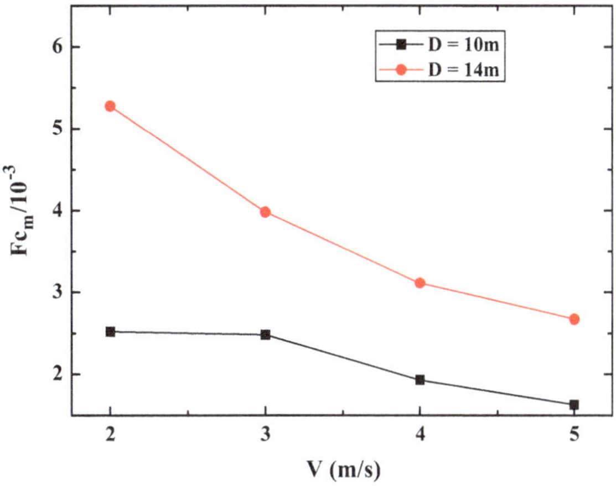

Fig. 4.7 Variation of mass transfer field synergy number with ACV

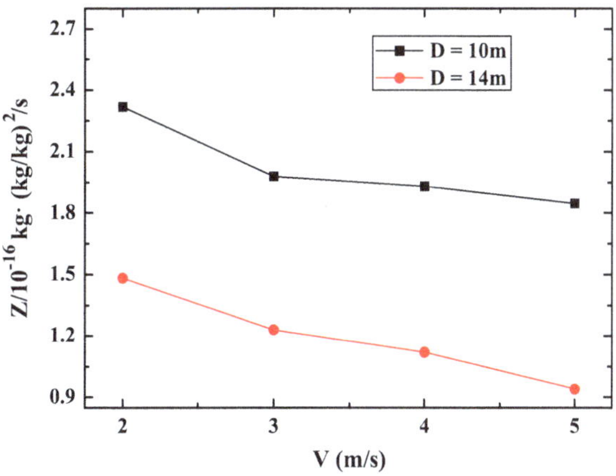

Fig. 4.8 Variation of mass entransy dissipation with ACV

assess the intensity. The decreased pollutant concentration decreases the concentration gradient and mass entransy dissipation function with increasing ACV, which enhances the overall decontamination rate. Meanwhile, mass transfer field synergy number grows in inverse proportion to pollutant concentration, so it decreases when the ACV increases. Accordingly, mass transfer field synergy number fails to evaluate the pollutant dilution ability of different ACVs. From Fig. 4.7, we can see that the pollutant dilution ability per unit volume is weak due to the decreased mass transfer field synergy number; in other words, the air utilization efficiency reduces. In conclusion, mass entransy dissipation function can be used to assess the intensity of dilution effect for the whole field domain, but mass transfer field synergy number can be used to assess the intensity of dilution effect per unit volume. Mass transfer field synergy number represents the convective mass transfer essential character more relative to mass entransy dissipation function. In the following sections, the relationship between mass transfer field synergy number and pollutant concentration will be explored by altering street width with ACV being kept constant. The study further strengthened the evidence linking mass transfer field synergy numbers with pollutant dilution ability hereinafter.

The increased ACV increases the Reynolds number, the overall Sherwood number, and the surface convective mass transfer coefficient.

4.4.1.3 Analysis of the Mass Field Synergy for Various Street Widths in 2D Model

The street canyon configuration has significant influence on the flow pattern and the pollutant dispersion from the traffic emission source. In order to reveal the influence of different street canyon configurations, the street width was set to 10.0, 14.0, 18.0, 22.0, and 26.0 m, respectively with the building height being fixed and the pollutant source intensity being kept constant, 5e−7 kg/m^3 s. In this section, ACV was set as 2 m/s for each street width. The relationship between mass transfer field synergy principle and pollutant concentration was thus presented as below.

Figures 4.9, 4.10, and 4.11 show the results of pollutant mass fraction, mass transfer field synergy number, and mass entransy dissipation, respectively. In Fig. 4.9, pollutant concentrations for various street widths are polar opposites with ACV being constant. Seen from Fig. 4.9, with the same ACV, the pollutant concentrations vary significantly with street width. Taking V = 2 m/s for example, the wider the street is, the lower its mean concentration of pollutant of the flow field is. In the meantime, it could be seen from Figs. 4.10 and 4.11 that when the ACV remains constant, the variation of field synergy number in the flow field is in accordance with that of mass entransy dissipation and pollutant concentration. When the street width increases, the pollutant concentration decreases, so does the

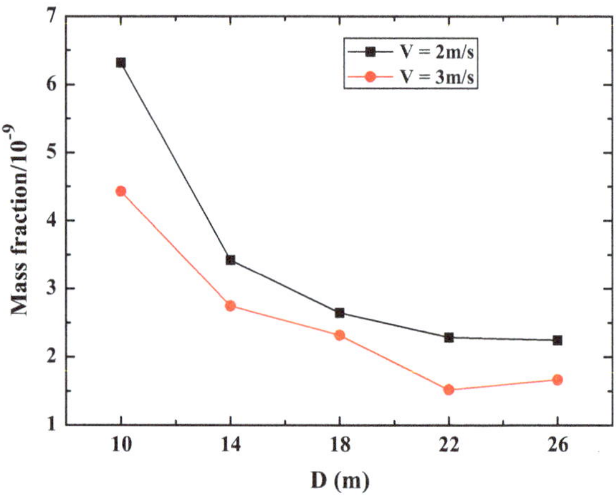

Fig. 4.9 Variation of mass entransy dissipation for various street widths

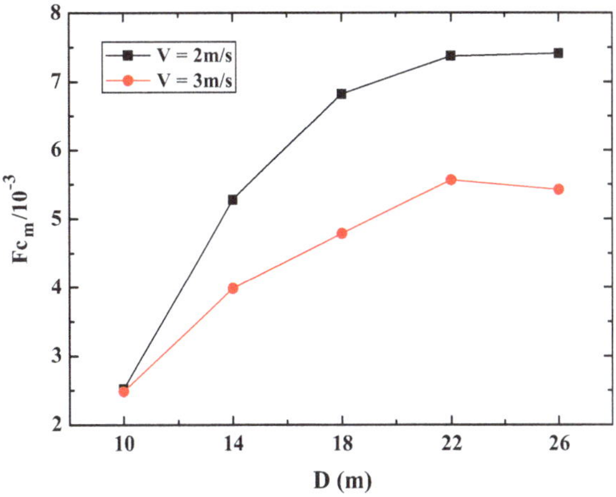

Fig. 4.10 Variation of mass transfer field synergy number for various street widths

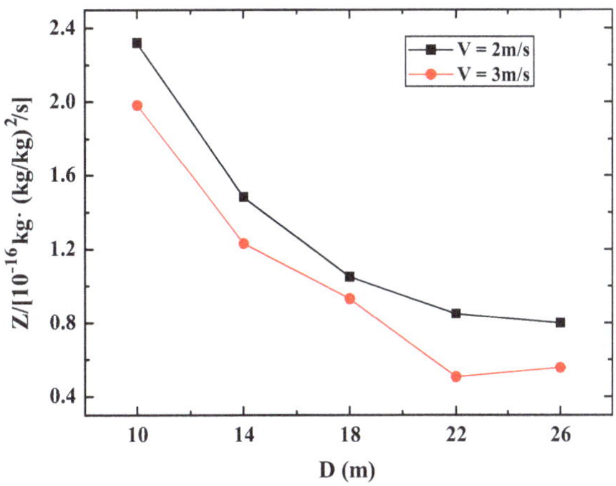

Fig. 4.11 Variation of mass entransy dissipation for various street widths

mass entransy dissipation and the mass transfer synergy number, which indicates that the convective mass transfer field influences the mass transfer performance of system. The higher the field synergy is, the better the performance of convective mass transfer is, and the stronger the pollutant dilution capacity is. In addition, since in the model D = 26 m where V = 3 m/s, the mean pollutant concentration in the field is greater than that of the model D = 24 m, the field synergy numbers are smaller than that of the model D = 24 m, which further indicates that the synergy angle of velocity distribution and component concentration gradient influence the convective mass transfer performance of the flow field.

Through the alteration of the boundary structure of flow field, the synergy angle of mass transfer field should increase. Namely, increasing the mass transfer synergy number could also increase the Sherwood number of the flow field, hence the intensity of convective mass transfer.

4.4.1.4 Further Explain the Mass Field Synergy Phenomenon in 2D Model

As aforementioned, when there exists the inlet characteristic velocity V = 3 m/s, the mean pollutant concentration of the street canyon model D = 26 m is higher than that D = 22 m. Therefore, through the analysis of the distributions of velocity vector and concentration gradient, we could explain the reason for this phenomenon.

Comparing Figs. 4.12 and 4.13, we could find that under the condition the inlet characteristic velocity is set as V = 3 m/s, when the height-width ratio is less than 1, increasing the street width on the contrary lowers the wind velocity in the street canyon. Combining the analysis with FSP, we could find that the field synergy number increases when the field synergy angles in the flow field of street canyon are roughly the same. Thus in the street canyon D = 22 m, the pollutant is much easier to be diluted by ambient crosswind, which makes the pollutant concentration even lower. The phenomenon that the velocity distribution in the street canyon

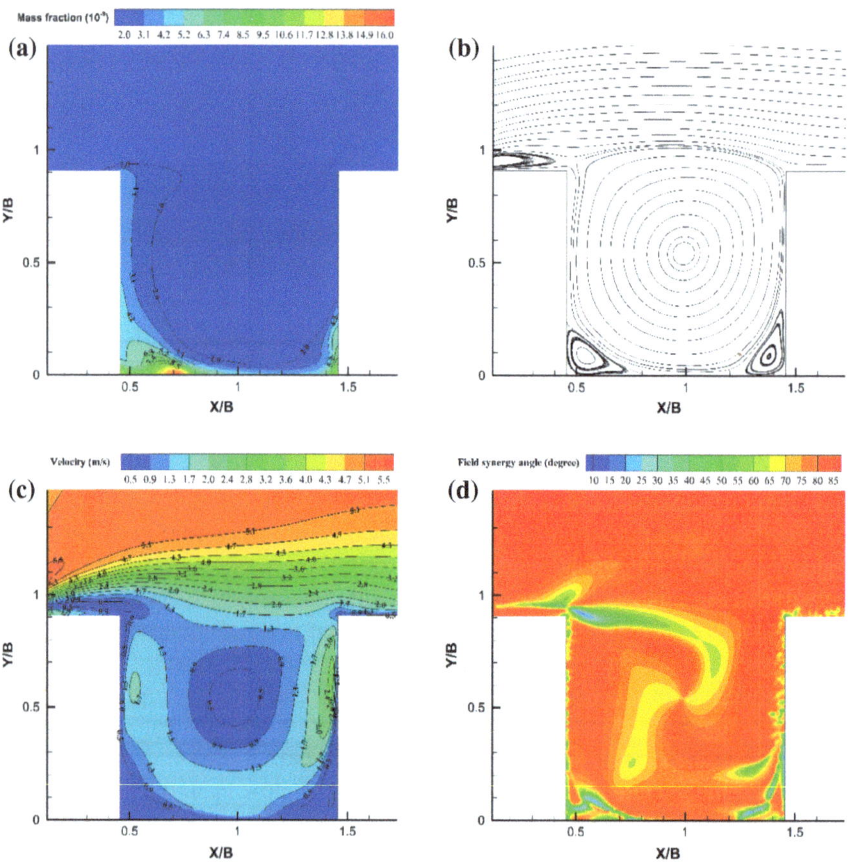

Fig. 4.12 Distribution in the first street canyon (U_{ref} = 3 m/s, D = 22 m): **a** normalized concentration distribution; **b** streamline; **c** velocity; **d** mass transfer field synergy angle

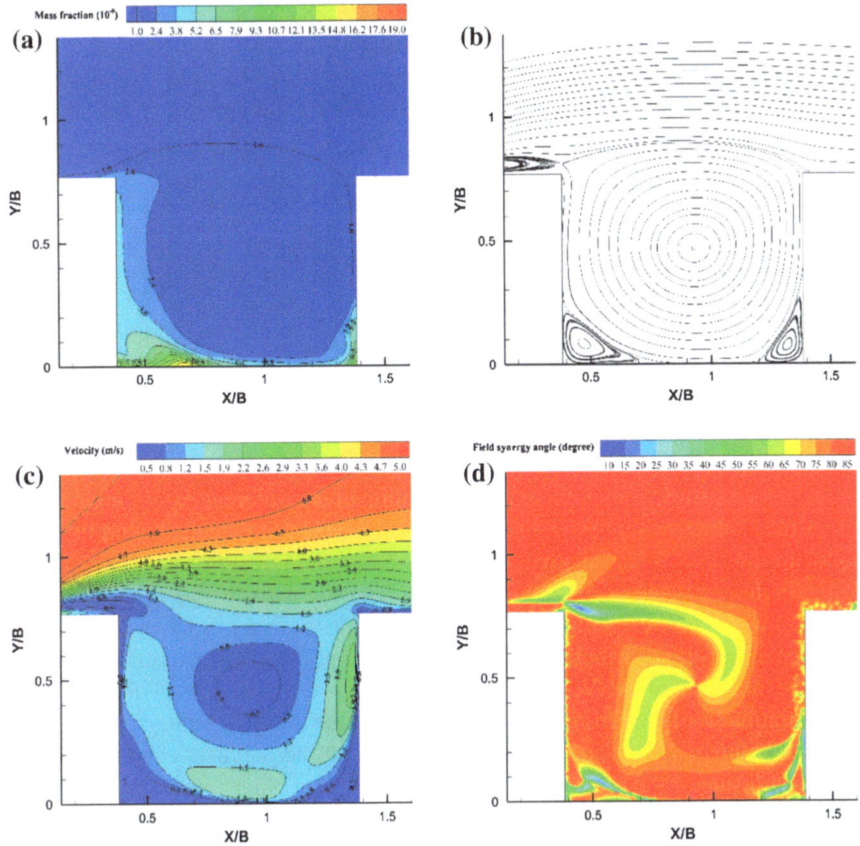

Fig. 4.13 Distribution in first street canyon (U_{ref} = 3 m/s, D = 18 m): **a** normalized concentration; **b** streamline; **c** velocity; **d** mass transfer field synergy angle

D = 22 m is higher than that when D = 26 m could be explained via the field synergy number from the velocity and the vector. The larger the synergy number is, the larger the fluid resistance is, and the increase of the turbulence energy in the flow field has increased the disturbance.

From this section, the relation between mass field synergy number and pollutant concentration becomes even more marked, higher mass field synergy number will cause lower pollutant concentration.

4.4.2 3D Model Analysis

According to the analysis of the validity and the correctness of the convective mass transfer FSP of the two-dimension planar model aforementioned, the similar case in the three-dimension model shall also be verified.

Via the establishment of two columns of buildings with the height 20 m, the width 10 m, and the length 40 m, perpendicular to the ambient crosswind, a street canyon between building columns was formed, with the pollutant in uniform distribution 0–2 m above the ground surface. Gradient wind appears at the inlet, the pollutant concentration source is set as $5e-7\,\text{kg/m}^3\,\text{s}$. In the whole wind tunnel flow field, the dispersion coefficient of CO is set as $2.88e-5\,\text{m}^2/\text{s}$.

4.4.2.1 Characteristics of the Flow Field for Various ACV in y = 12*B* Plane

Four tests were designed for each ambient wind speed to investigate the influence of thermal effects with different ambient wind speeds. The street canyon configuration presenting two columns of parallel buildings with aspect ratio $H/D = 1$ is investigated.

Figure 4.14 shows the pollutant concentration under different ACV ranging from 2 to 5 m/s. Numerous investigations have been devoted to elucidate wind flow and pollutant transport in urban street canyons through the three-dimension mode. The region in the x–z plane (y = 12*B*) was chosen. The comparison among the simulated results of different ACV indicates that the wind velocity is a key factor determining the intensity of the distribution in the street canyons. The concentrations decrease with the increase of ACV. It is noticeable that the maximum of the pollutant concentration has fallen from 26 to 13 (10^{-9}) because the ACV improves the ability to dilute pollutants. The calculated results show that the pollutant concentration on the leeward side of the upstream building is much higher than that on the windward side of the downstream building. From Fig. 4.15, the mass transfer field synergy angle on the windward side was obviously lager than that on the leeward side. So as a result, the synergic mechanism of velocity vector and pollutant concentration gradient can promote the dilution and diffusion of pollutant. Despite similar field synergy angles distribution for various ACV, the local velocity have a big effect on the mass field synergy number which is verified in 2D model. The phenomenon that the pollutant concentration declines with increasing ACV increases can be explained from the mass transfer synergy point.

Figure 4.16 shows that wind velocity field represents how things are moving at every point of the plane. There are two regions where the wind velocity is less than 0.5 m/s in the x–z plane. One is at the bottom of street canyon and another is in the vortex center which is seen from Fig. 4.17. It means that the regions are bad for the air circulation. The magnitude of wind velocity can significantly affect mass transfer synergy level according to field synergy principle. With the ambient crosswind speed increasing, the low wind speed region gradually shrinks, so does the level of the concentrations of whole flow field.

Taking the direct sunshine into account, we found that both the building facade and the ground surface are heated simultaneously. The flow which is used to simulate the process of the dilution of pollutants consisting of co and air, and it was

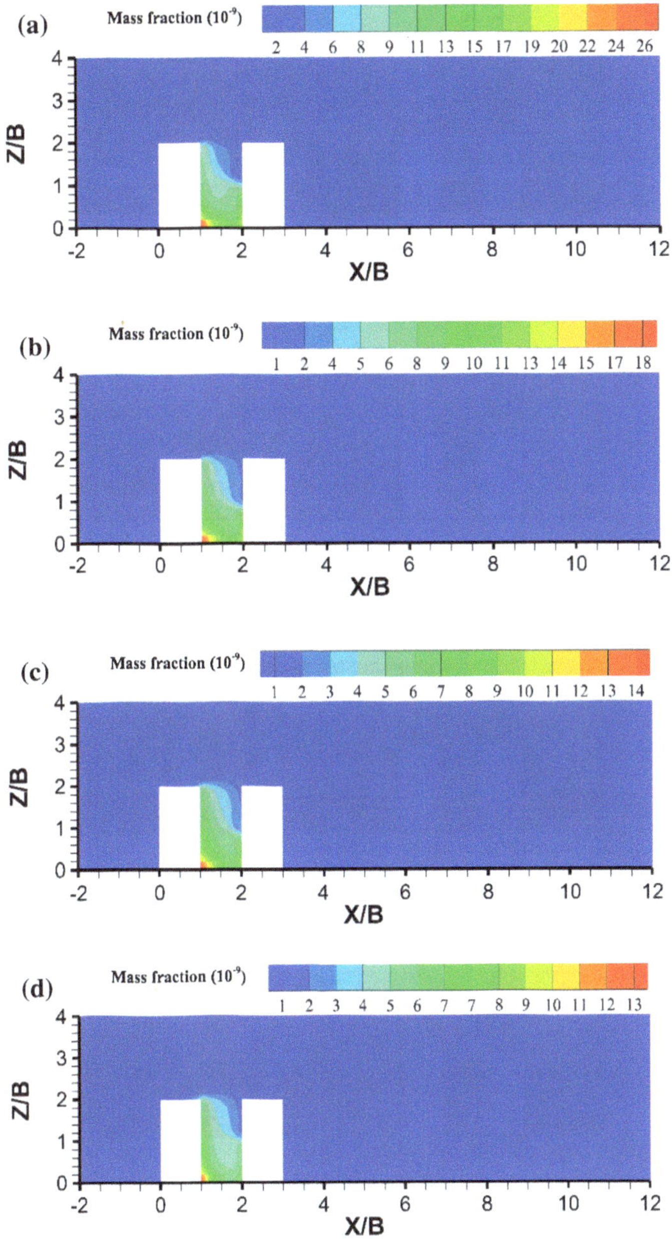

Fig. 4.14 Distribution of pollutant concentration in $y = 12B$: **a** $v = 2$ m/s, **b** $v = 3$ m/s, **c** $v = 4$ m/s, **d** $v = 5$ m/s

set as incompressible turbulent inert flow. When solar radiation heats the building facade or the ground, the air density will change with increasing temperature. The buoyancy forces driven by density variations contribute to the fluid motion on

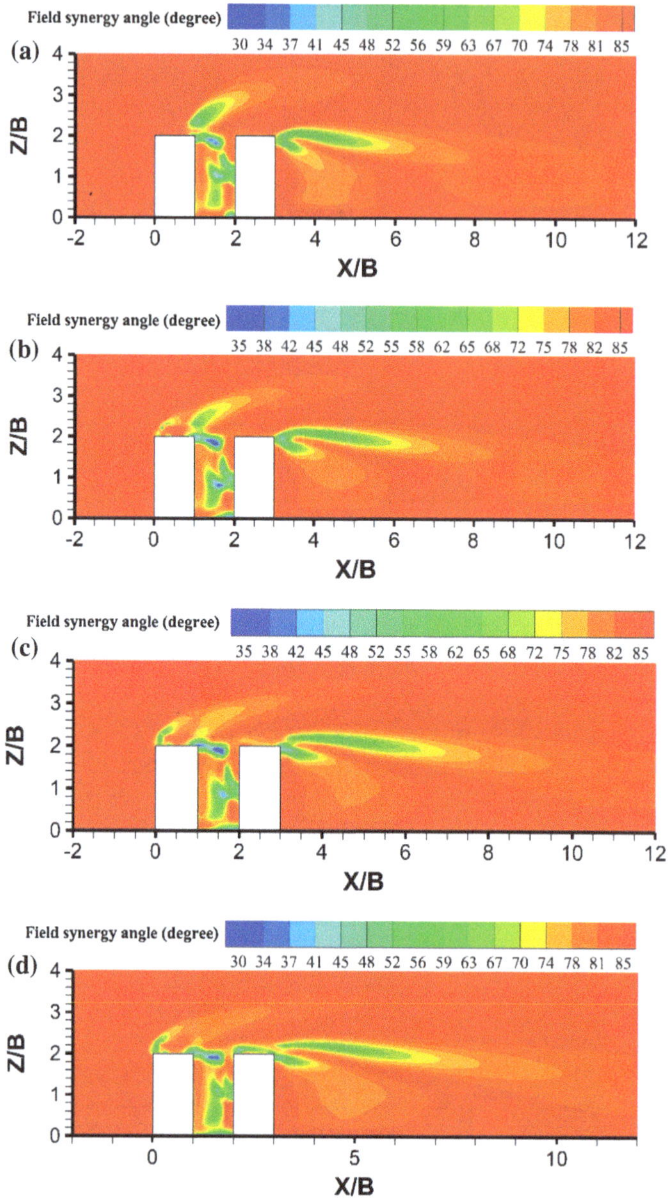

Fig. 4.15 Distribution of mass transfer field synergy angle in the x–z plane ($y = 12B$): **a** $v = 2$ m/s, **b** $v = 3$ m/s, **c** $v = 4$ m/s, **d** $v = 5$ m/s

building height according to Boussinesq's hypothesis. The relationship between fluid temperature and Bulk Richardson Number affect the distribution of pollutant concentrations. Once the wall temperature reaches a certain value, the thermal effect

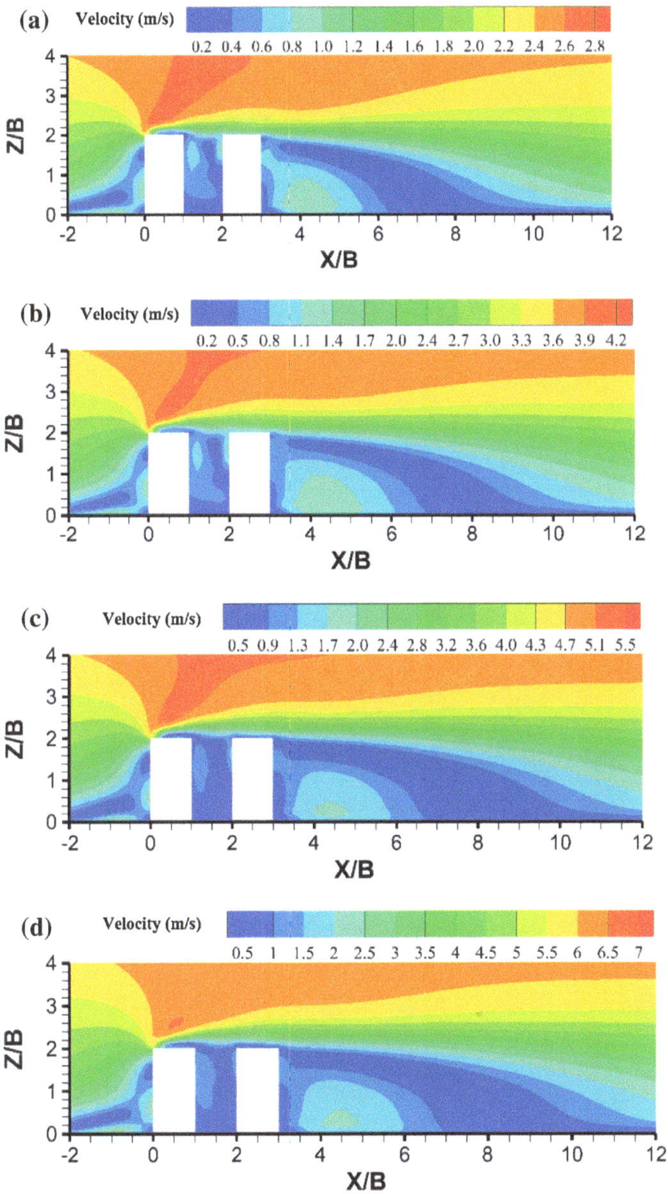

Fig. 4.16 Distribution of velocity in $y = 12B$: **a** $v = 2$ m/s, **b** $v = 3$ m/s, **c** $v = 4$ m/s, **d** $v = 5$ m/s

is more obvious as the temperature difference increases between the fluid and the building wall. Figures 4.14 and 4.18 show that temperature and concentrations of pollutant have similar distributions in the street canyon. This is because the force to dilute the outdoor air pollutants is driven by the fluid temperature. That the regions

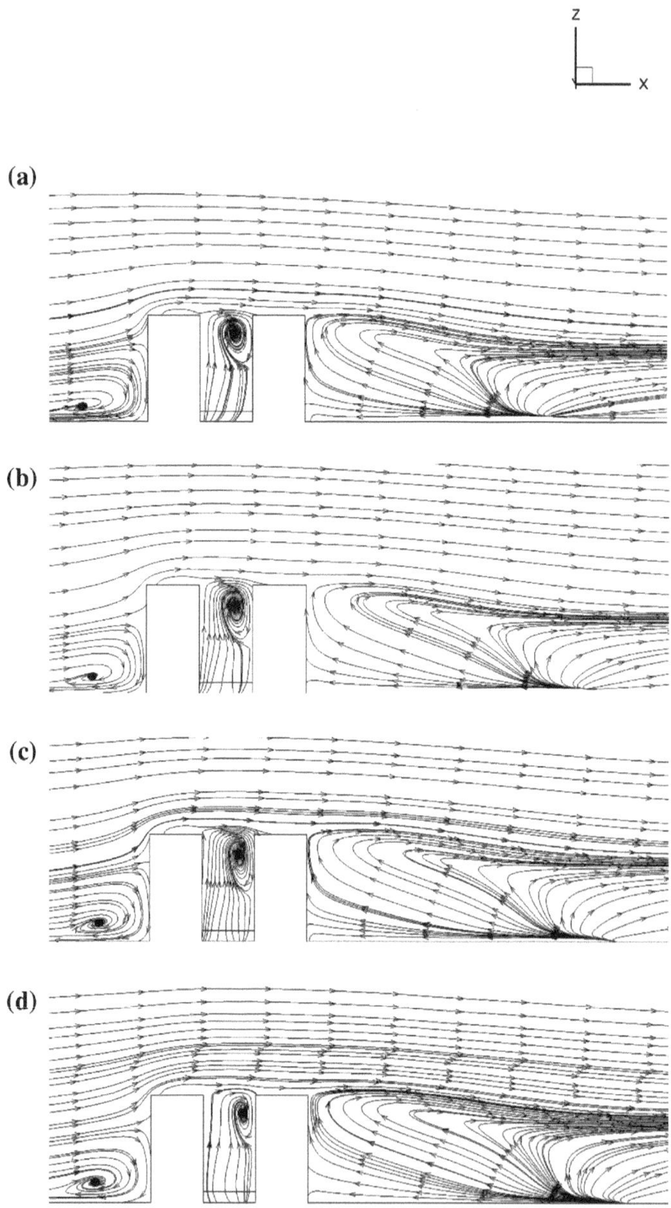

Fig. 4.17 Flow field patterns under different ambient wind speeds in the street canyon with H/D ratio of 2 in $y = 12B$: **a** $v = 2$ m/s, **b** $v = 3$ m/s, **c** $v = 4$ m/s, **d** $v = 5$ m/s

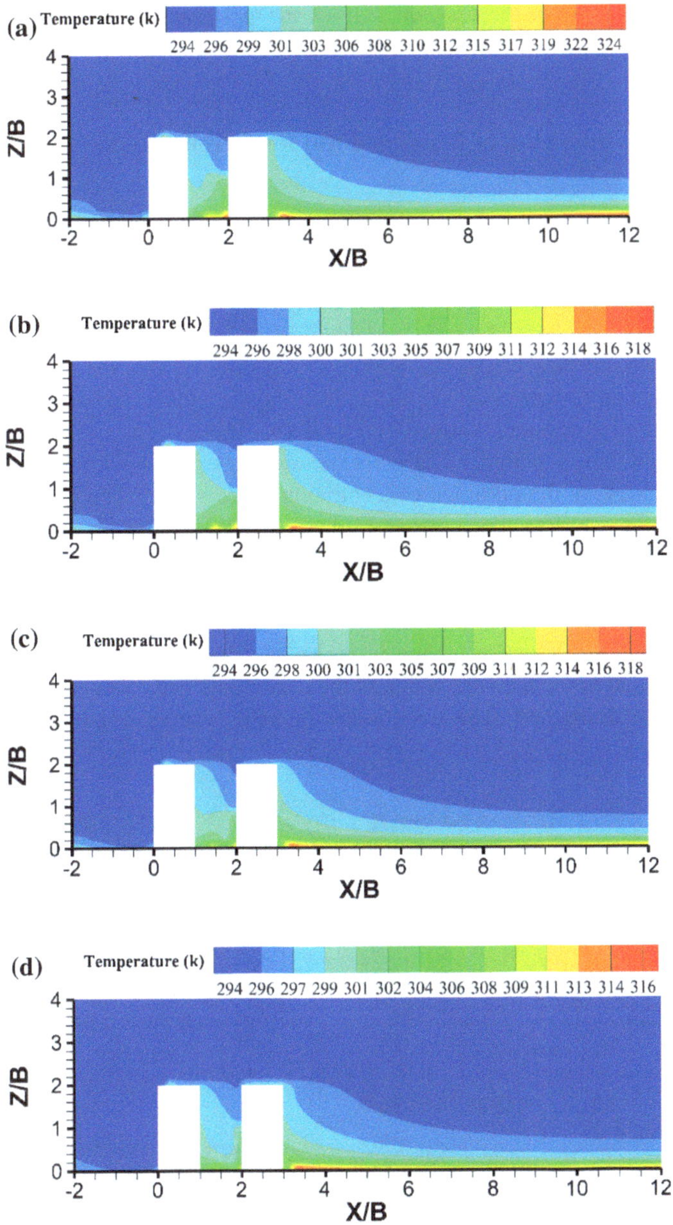

Fig. 4.18 Distribution of temperature in $y = 12B$: **a** $v = 2$ m/s, **b** $v = 3$ m/s, **c** $v = 4$ m/s, **d** $v = 5$ m/s

gathering pollutant on the leeward side of the upstream building get more heat results in the increase of the temperature and concentrations.

4.4.2.2 Characteristics of the Flow Field for Various ACV in $z = 2B$ Plane

In Figs. 4.19, 4.20, 4.21, and 4.22, it can be seen that the fields for pollutant concentration and velocity components correlate with the plane x–y ($z = 2B$). These figures analyzed, it is noticeable that when the free stream impacts against a building perpendicularly, the stream would diverge into two sub-streams in front of the buildings and flow across the building. It is obvious that the average concentration values on leeward wall are higher than those on the windward wall based on the concentration contour shown in Fig. 4.19. The extreme value of pollutant concentration has experienced an increase–decrease trend with the increase of ACV. The cause of the pollutant concentration decrease has been explained earlier in this article, whereas the increase is due to turbulent disturbance on building height accordingly. The increase of ACV makes the pollutant transport much easier.

Figure 4.20 shows the velocity components on the plane x–y ($z = 2B$). It is obvious that the turbulent motion due to the increase of ACV extends to both ends in the y direction. Meanwhile, the increase of velocity causes a larger velocity gradient normal to the leeward wall and the average velocity values beside leeward wall increase drastically. The distribution of field synergy angle is presented in

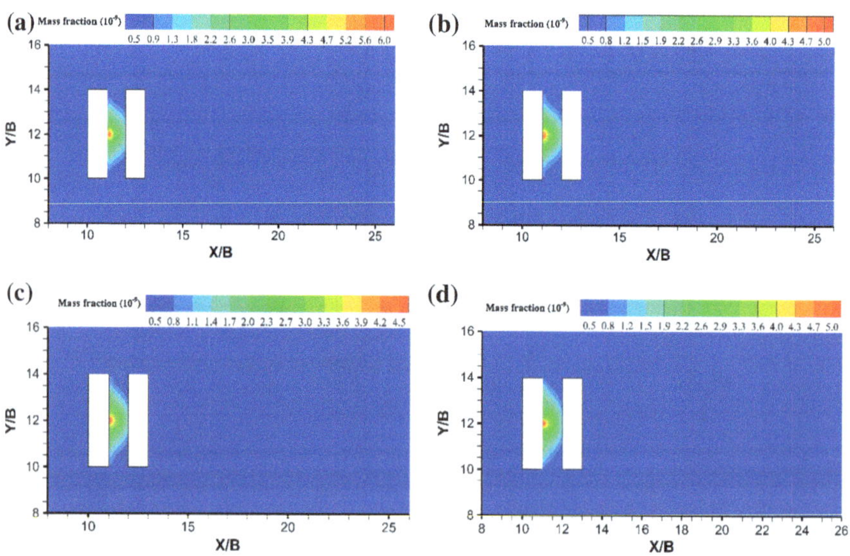

Fig. 4.19 Distribution of pollutant concentration in the x–y plane ($z = 2B$): **a** $v = 2$ m/s, **b** $v = 3$ m/s, **c** $v = 4$ m/s, **d** $v = 5$ m/s

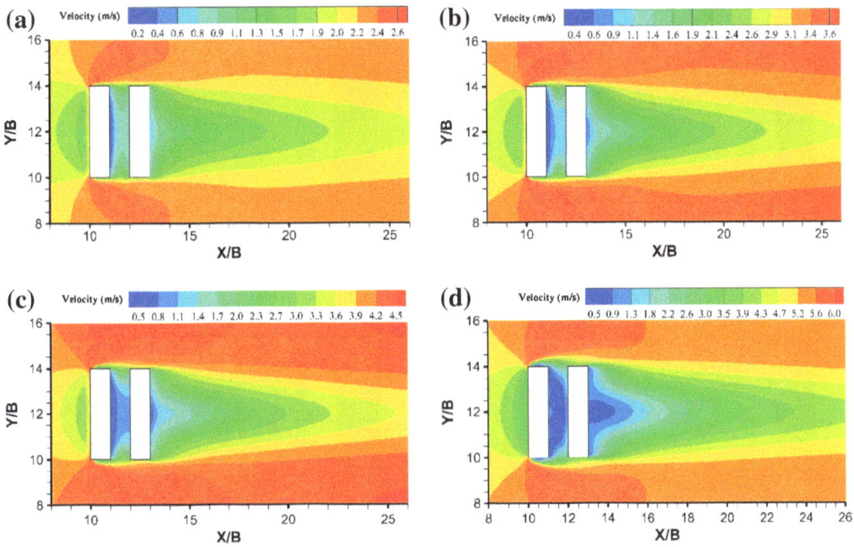

Fig. 4.20 Distribution of velocity in the *x*–*y* plane ($z = 2B$): **a** $v = 2$ m/s, **b** $v = 3$ m/s, **c** $v = 4$ m/s, **d** $v = 5$ m/s

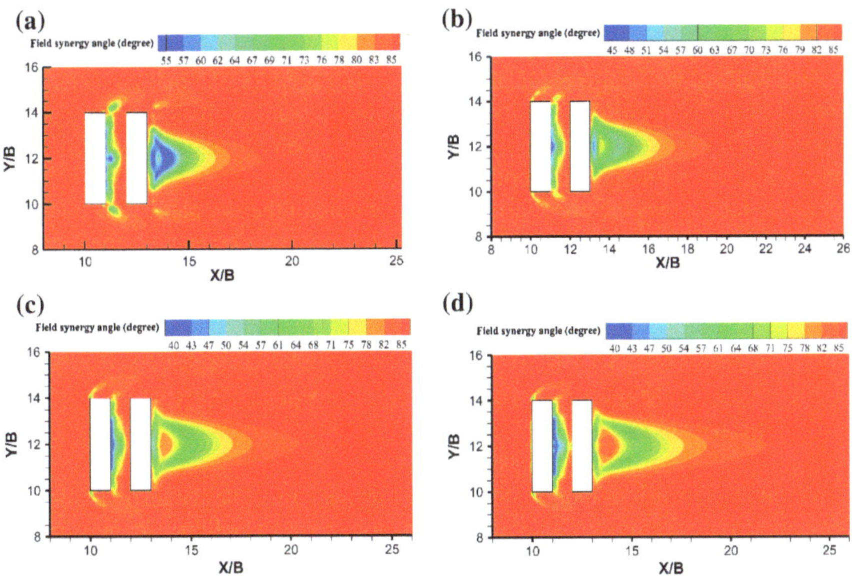

Fig. 4.21 Distribution of field synergy angle in the *x*–*y* plane ($z = 2B$): **a** $v = 2$ m/s, **b** $v = 3$ m/s, **c** $v = 4$ m/s, **d** $v = 5$ m/s

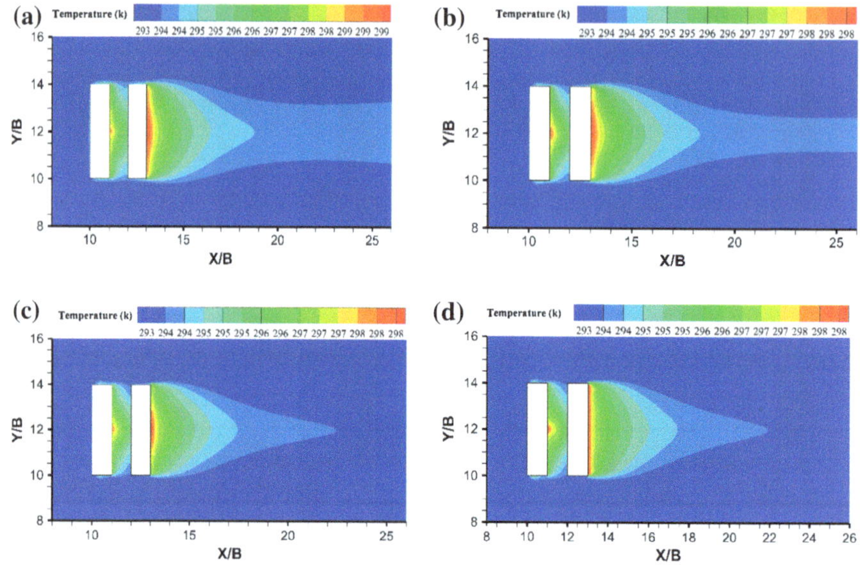

Fig. 4.22 Distribution of temperature in the x–y plane ($z = 2B$): **a** $v = 2$ m/s, **b** $v = 3$ m/s, **c** $v = 4$ m/s, **d** $v = 5$ m/s

Fig. 4.21, the synergy angle between the velocity field and concentration field embodies the mass transfer capacity according to FSP. Besides, the contour profile of the distribution of pollutant concentration appears as a circular arc centered in the leeward wall, which is also explicable via FSP. In Fig. 4.21, the mass transfer field synergy angle is nearly 90° at the edge of the leeward wall, it means that the velocity vector is perpendicular to concentration gradient which is bad for mass transfer and the dilution of the pollutants.

Figure 4.22 shows the distribution of temperature in the x–y plane ($z = 2B$). The distribution of temperature is not very different from each other with the increase of ACV, and the temperature has little contribution to the mass transfer in the x–y plane because buoyancy force is normal to the x–y plane.

4.4.2.3 Analysis of the Mass Field Synergy for Various ACV in 3D Model

In the three-dimension model, the mass transfer principle is also applicable. Comparing the relationship among the variation trend of pollutant concentration, the mass transfer synergy number and the mass entransy dissipation in Figs. 4.23 and 4.24, we could find that when the ACV increases, the pollutant concentration in the flow field decreases, the mass transfer synergy number decreases, and at the same time the mass transfer potential capacity dissipation coefficient decreases, which is identical with the variation trend of two-dimension model. The mass

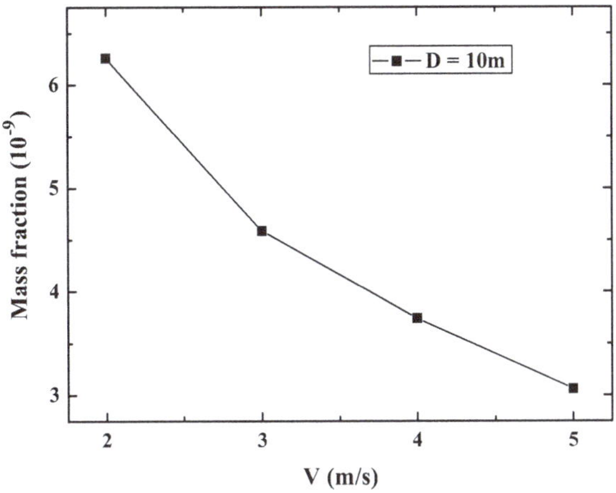

Fig. 4.23 Variation of pollutant mass fraction for various street width in 3D model

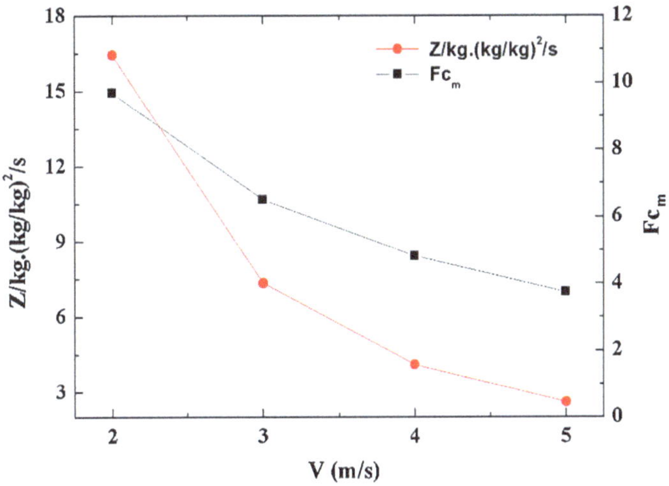

Fig. 4.24 Variation of for mass entransy dissipation Z and mass transfer field synergy numbers Fc_m various street width in 3D model

transfer synergy number indicates the pollutant dilution of air volume per unit. Due to the increase of air mass flow, the purifying capacity in the flow field is enhanced, which could be derived from the results of mass entransy dissipation. The lower the potential capacity of dissipation coefficient is, the lower the pollutant concentration of the flow field is. Meanwhile, similar to the results of two-dimension planar model, when the velocity increases, the decrease range of pollutant concentration will decrease. The reason is that the mass transfer synergy number decreases.

Therefore, increasing the inlet velocity could enhance the purification effect, which however, is not economical. Instead, the modification of block structure and the optimization of topo-flow field could also enhance the pollutant dilution.

4.5 Conclusions

In this chapter, we further extended the heat transfer field synergy principle to mass transfer field synergy. Then we analyzed the synergy angle of velocity with concentration gradient, based upon which, we also revealed the relationship between synergy angle and pollutant concentration. The following conclusions are drawn by the findings:

(1) In order to solve the contradiction that the traditional dilution effect is only a qualitative, mass transfer field synergy number and entransy dissipation are taken as the quantitative calculations to assess the intensity of dilution effect. The relationship between convection mass transfer performance and the effect of mass transfer field synergy is verified via 2D and 3D model for various ACV and street width. With ACV increasing, the pollutant concentration and mass entransy dissipation, and mass transfer field synergy number fall simultaneously. When street width increases, mass entransy dissipation and mass transfer field synergy number grow in inverse proportion to pollutant concentration.
(2) Mass transfer field synergy number can represent the convective mass transfer essential character. In the ambient wind passing through street canyon processes, the overall street canyon decontamination rate can be enhanced by increasing ACV, but pollutant dilution ability per unit volume is weak due to the decreased mass transfer field synergy number.
(3) Mass transfer field synergy angle and flow field velocity as critical component of mass transfer field synergy number can have effect on the ability of pollutant dispersion. With synergy angle being constant, increased field velocity can decrease pollutant concentration. Similarly, decreased synergy angle can decrease pollutant concentration.

References

1. Z.Y. Guo, D.Y. Li, B.X. Wang. A novel concept for convective heat transfer enhancement. International Journal of Heat & Mass Transfer. 41 (1998) 2221–5.
2. Z.Y. Guo, W.Q. Tao, R.K. Shah. The field synergy (coordination) principle and its applications in enhancing single phase convective heat transfer. International Journal of Heat & Mass Transfer. 48 (2005) 1797–807.
3. J.A. Meng, X.G. Liang, Z.X. Li. Field synergy optimization and enhanced heat transfer by multi-longitudinal vortexes flow in tube. International Journal of Heat & Mass Transfer. 48 (2005) 3331–7.

4. W.Q. Tao, Y.L. He, Z.G. Qu, Y.P. Cheng. Applications of the field synergy principle in developing new type heat transfer enhanced surfaces. Journal of Enhanced Heat Transfer. 11 (2004) 435–52.
5. W.Q. Tao, Y.L. He, Q.W. Wang, Z.G. Qu, F.Q. Song. A unified analysis on enhancing single phase convective heat transfer with field synergy principle. International Journal of Heat & Mass Transfer. 45 (2002) 4871–9.
6. M. Zeng, W.Q. Tao, M. Zeng, W.Q. Tao. Numerical Verification of the Field Synergy Principle for Turbulent Flow. Journal of Enhanced Heat Transfer. 11 (2004) 453–60.
7. W.Q. Tao, W.W. Jin, Y.L. He, Z.G. Qu, C.C. Zhang. Optimum Design of Two-Row Slotted Fin Surface with X-Shape Strip Arrangement Positioned by "Front Coarse and Rear Dense" Principle, Part I: Physical/Mathematical Models and Numerical Methods. Numerical Heat Transfer Applications. 17 (2006) 731–49.
8. Y. Chen, X. Deng, X. Ding. CONVECTION HEAT TRANSFER OF CONVERGING-DIVERGING TUBE(I)MECHANISM OF FIELD SYNERGY PRINCIPLE. Journal of Chemical Industry & Engineering. (2004).
9. W.U. Yan-Yang, W.H. Yang, Y.U. Jiu-Yang, L. Wei, J.Y. Gao, L.J. Liu, et al. Numerical Analysis on Convection Heat Transfer in a Spirally Fluted and Field Synergy Principle Analysis. Petro-Chemical Equipment. 308–310 (2011) 1410–5.
10. W.U. Ming. Extension and application of field synergy principle to pulse tube refrigerator. Journal of Engineering Thermophysics. 23 (2002) 488–90.
11. J. Guo, M. Xu, C. Lin. The application of field synergy number in shell-and-tube heat exchanger optimization design. Applied Energy. 86 (2009) 2079–87.
12. W. Liu, K. Yang. Mechanism and numerical analysis of heat transfer enhancement in the core flow along a tube. Science in China Series E: Technological Sciences. 51 (2008) 1195–202.
13. A. Nakayama, F. Kuwahara, W. Liu. A macroscopic model for countercurrent bioheat transfer in a circulatory system. Journal of Porous Media. 12 (2009).
14. Z.F. Huang, A. Nakayama, K. Yang, C. Yang, W. Liu. Enhancing heat transfer in the core flow by using porous medium insert in a tube. International Journal of Heat and Mass Transfer. 53 (2010) 1164–74.
15. T.Z. Ming, Y. Zheng, J. Liu, C. Liu, W. Liu, S.Y. Huang. Heat transfer enhancement by filling metal porous medium in central area of tubes. Journal of the Energy Institute. 83 (2010) 17–24.
16. C. Yang, A. Nakayama, W. Liu. Heat transfer performance assessment for forced convection in a tube partially filled with a porous medium. International Journal of Thermal Sciences. 54 (2012) 98–108.
17. J. Guo, Y. Yan, W. Liu, F. Jiang, A. Fan. Enhancement of laminar convective heat transfer relying on excitation of transverse secondary swirl flow. International Journal of Thermal Sciences. 87 (2015) 199–206.
18. J.J. Liu, Z.C. Liu, W. Liu. 3D numerical study on shell side heat transfer and flow characteristics of rod-baffle heat exchangers with spirally corrugated tubes. International Journal of Thermal Sciences. 89 (2015) 34–42.
19. W. Liu, Z. Liu, Y. Wang, S. Huang. Flow mechanism and heat transfer enhancement in longitudinal-flow tube bundle of shell-and-tube heat exchanger. Science in China Series E: Technological Sciences. 52 (2009) 2952–9.
20. Y. Wang, Z. Liu, S. Huang, W. Liu, W. Li. Experimental investigation of shell-and-tube heat exchanger with a new type of baffles. Heat and Mass Transfer. 47 (2011) 833–9.
21. J. Yang, W. Liu. Numerical investigation on a novel shell-and-tube heat exchanger with plate baffles and experimental validation. Energy Conversion and Management. 101 (2015) 689–96.
22. J. Yang, L. Ma, J. Bock, A.M. Jacobi, W. Liu. A comparison of four numerical modeling approaches for enhanced shell-and-tube heat exchangers with experimental validation. Applied Thermal Engineering. 65 (2014) 369–83.
23. J. Guo, A. Fan, X. Zhang, W. Liu. A numerical study on heat transfer and friction factor characteristics of laminar flow in a circular tube fitted with center-cleared twisted tape. International Journal of Thermal Sciences. 50 (2011) 1263–70.

24. P. Li, Z. Liu, W. Liu, G. Chen. Numerical study on heat transfer enhancement characteristics of tube inserted with centrally hollow narrow twisted tapes. International Journal of Heat and Mass Transfer. 88 (2015) 481–91.
25. P. Liu, N. Zheng, F. Shan, Z. Liu, W. Liu. Numerical study on characteristics of heat transfer and friction factor in a circular tube with central slant rods. International Journal of Heat and Mass Transfer. 99 (2016) 268–82.
26. X. Zhang, Z. Liu, W. Liu. Numerical studies on heat transfer and flow characteristics for laminar flow in a tube with multiple regularly spaced twisted tapes. International Journal of Thermal Sciences. 58 (2012) 157–67.
27. N. Zheng, P. Liu, F. Shan, Z. Liu, W. Liu. Effects of rib arrangements on the flow pattern and heat transfer in an internally ribbed heat exchanger tube. International Journal of Thermal Sciences. 101 (2016) 93–105.
28. N. Zheng, W. Liu, Z. Liu, P. Liu, F. Shan. A numerical study on heat transfer enhancement and the flow structure in a heat exchanger tube with discrete double inclined ribs. Applied Thermal Engineering. 90 (2015) 232–41.
29. Y. Wang, B. Zhou, Z. Liu, Z. Tu, W. Liu. Numerical study and performance analyses of the mini-channel with discrete double-inclined ribs. International Journal of Heat and Mass Transfer. 78 (2014) 498–505.
30. N. Zheng, P. Liu, F. Shan, Z. Liu, W. Liu. Numerical investigations of the thermal-hydraulic performance in a rib-grooved heat exchanger tube based on entropy generation analysis. Applied Thermal Engineering. 99 (2016) 1071–85.
31. H. Jia, W. Liu, Z. Liu. Enhancing convective heat transfer based on minimum power consumption principle. Chemical Engineering Science. 69 (2012) 225–30.
32. H. Jia, Z.C. Liu, W. Liu, A. Nakayama. Convective heat transfer optimization based on minimum entransy dissipation in the circular tube. International Journal of Heat and Mass Transfer. 73 (2014) 124–9.
33. J. Yang, S.-R. Oh, W. Liu. Optimization of shell-and-tube heat exchangers using a general design approach motivated by constructal theory. International Journal of Heat and Mass Transfer. 77 (2014) 1144–54.
34. Y. Ge, Z. Liu, W. Liu, G. Chen. Active optimization design theory and method for heat transfer unit and its application on shape design of cylinder in convective heat transfer. International Journal of Heat and Mass Transfer. 90 (2015) 702–9.
35. J. Wang, Z. Liu, F. Yuan, W. Liu, G. Chen. Convective heat transfer optimization in a circular tube based on local exergy destruction minimization. International Journal of Heat and Mass Transfer. 90 (2015) 49–57.
36. Y. Ge, Z. Liu, W. Liu. Multi-objective genetic optimization of the heat transfer for tube inserted with porous media. International Journal of Heat and Mass Transfer. 101 (2016) 981–7.
37. J. Wang, Z. Liu, W. Liu. Evaluation of convective heat transfer in a tube based on local exergy destruction rate. Science China Technological Sciences. 59 (2016) 1494–506.
38. W. Liu, Z.C. Liu, H. Jia, A.W. Fan, A. Nakayama. Entransy expression of the second law of thermodynamics and its application to optimization in heat transfer process. International Journal of Heat and Mass Transfer. 54 (2011) 3049–59.
39. R. Long, W. Liu. Ecological optimization and coefficient of performance bounds of general refrigerators. Physica A: Statistical Mechanics and its Applications. 443 (2016) 14–21.
40. R. Long, W. Liu. Unified trade-off optimization for general heat devices with nonisothermal processes. Phys Rev E Stat Nonlin Soft Matter Phys. 91 (2015) 042127.
41. R. Long, B. Li, Z. Liu, W. Liu. Multi-objective optimization of a continuous thermally regenerative electrochemical cycle for waste heat recovery. Energy. 93 (2015) 1022–9.
42. R. Long, Y.J. Bao, X.M. Huang, W. Liu. Exergy analysis and working fluid selection of organic Rankine cycle for low grade waste heat recovery. Energy. 73 (2014) 475–83.
43. R. Long, B. Li, Z. Liu, W. Liu. Performance analysis of a solar-powered electrochemical refrigerator. Chemical Engineering Journal. 284 (2016) 325–32.
44. R. Long, B. Li, Z. Liu, W. Liu. Performance analysis of a solar-powered solid state heat engine for electricity generation. Energy. 93 (2015) 165–72.

45. R. Long, W. Liu. Performance of quantum Otto refrigerators with squeezing. Phys Rev E Stat Nonlin Soft Matter Phys. 91 (2015) 062137.
46. R. Long, W. Liu. Ecological optimization for general heat engines. Physica A: Statistical Mechanics and its Applications. 434 (2015) 232–9.
47. R. Long, Z. Liu, W. Liu. Performance optimization of minimally nonlinear irreversible heat engines and refrigerators under a trade-off figure of merit. Phys Rev E Stat Nonlin Soft Matter Phys. 89 (2014) 062119.
48. R. Long, W. Liu. Coefficient of performance and its bounds for general refrigerators with nonisothermal processes. Journal of Physics A: Mathematical and Theoretical. 47 (2014) 325002.
49. R. Long, B. Li, Z. Liu, W. Liu. Performance analysis of a thermally regenerative electrochemical cycle for harvesting waste heat. Energy. 87 (2015) 463–9.
50. W. Liu, Z. Liu, Z. Guo. Physical quantity synergy in laminar flow field of convective heat transfer and analysis of heat transfer enhancement. Science Bulletin. 54 (2009) 3579–86.
51. L. Wei, L. Zhichun, M. Tingzhen, G. Zengyuan. Physical quantity synergy in laminar flow field and its application in heat transfer enhancement. International Journal of Heat and Mass Transfer. 52 (2009) 4669–72.
52. W. Liu, Z. Liu, S. Huang. Physical quantity synergy in the field of turbulent heat transfer and its analysis for heat transfer enhancement. Chinese Science Bulletin. 55 (2010) 2589–97.
53. W. Liu, Z. Liu, L. Ma. Application of a multi-field synergy principle in the performance evaluation of convective heat transfer enhancement in a tube. Chinese Science Bulletin. 57 (2012) 1600–7.
54. L. Ma, J. Yang, W. Liu, X. Zhang. Physical quantity synergy analysis and efficiency evaluation criterion of heat transfer enhancement. International Journal of Thermal Sciences. 80 (2014) 23–32.
55. J. Yang, L. Ma, J. Liu, W. Liu. Thermal–hydraulic performance of a novel shell-and-tube oil cooler with multi-fields synergy analysis. International Journal of Heat and Mass Transfer. 77 (2014) 928–39.
56. N. Zheng, P. Liu, F. Shan, Z. Liu, W. Liu. Heat transfer enhancement in a novel internally grooved tube by generating longitudinal swirl flows with multi-vortexes. Applied Thermal Engineering. 95 (2016) 421–32.
57. F. Shan, Z. Liu, W. Liu, Y. Tsuji. On flow structures associated with large wall mass transfer coefficients in orifice flows. International Journal of Heat and Mass Transfer. 102 (2016) 1–9.
58. F. Shan, Z. Liu, W. Liu, Y. Tsuji. Effects of the orifice to pipe diameter ratio on orifice flows. Chemical Engineering Science. 152 (2016) 497–506.
59. B. Li, R. Long, Z. Liu, W. Liu. Performance analysis of a thermally regenerative electrochemical refrigerator. Energy. 112 (2016) 43–51.
60. Q. Chen, J. Ren, Z. Guo. Field synergy analysis and optimization of decontamination ventilation designs. International Journal of Heat & Mass Transfer. 51 (2008) 873–81.
61. L.F. Wei, L. Qiu. Study of Field Synergy Theory on the Performance of Mass Transfer in Dehumidifying Air-conditioner. Refrigeration Air Conditioning & Electric Power Machinery. (2007).
62. Q. Chen, J.A. Meng. Field synergy analysis and optimization of the convective mass transfer in photocatalytic oxidation reactors. International Journal of Heat & Mass Transfer. 51 (2008) 2863–70.
63. Q. Chen, Z. Qin. Simulation on field synergy enhancement for convective mass transfer in helical tube. CIESC Journal. 49 (2012) 1195.
64. M. Schatzmann. Quality assurance and improvement of microscale meteorological models. International Journal of Environment & Pollution. 44 (2011) 139–46.

Chapter 5
A Solar Chimney with an Inverted U-Type Cooling Tower to Mitigate Urban Air Pollution

5.1 Introduction

The idea of solar chimney power plant (SCPP) was first put forward by Schlaich et al. [1]. It is based on the utilization of the air density decrease with increasing temperature. The air is heated in a solar collector, then it rises inside a chimney driven by buoyancy, and it drives turbines to generate electricity. In 1983, the world's first SCPP was built in Manzanares, Spain. This experimental SCPP with 194.6 m chimney height and 5.08 m radius was fully tested and validated till 1989. The relevant experimental results and a scientific description were given by Haff et al. [2, 3]. After that, more and more researchers engaged in the research of SCPP [4–13]. Some researchers also have proposed a series of novel SCPP systems [14–18]. However, it is worth mentioning that most researchers are more focused on how to improve the efficiency of the SC power generation.

Cao et al. [19] proposed a solar-assisted large-scale cleaning system for air pollution. The system consists of a large-scale solar collector with the radius of 2500 m, and a chimney with the height of 500 m. There is a filter bank placed near the entrance of the chimney, thus the $PM_{2.5}$ and larger particulate matter is separated from the air. Zhou et al. [20] proposed high SCs to drive the warm air containing haze up to higher altitude and enhance the dispersion of dense haze. They made creative use of urban heat island instead of a vast and expensive solar collector to provide warm air. Besides, Ming et al. [21] also suggested that the SC technology is able to transfer heat from the Earth surface to the upper layers of the troposphere, thus could cool down the Earth and combat climate change. Based on these ideas, it seems that the application of SC is a feasible approach to control the air pollution.

A majority of previous SC studies have focused on the indoor ventilation, few studies on the outdoor ventilation. More importantly, when SC is used in outdoor ventilation, the size should be very large which would raise some economic and engineering problem. Meanwhile, it is worth noting that the thermal airflow with a

T. Ming et al., *Pollutant Dispersion in Built Environment*,
DOI 10.1007/978-981-10-3821-1_5

high temperature flows out chimney outlet can not directly improve the air conditions in the spectrum of human activity, so the efficiency of ventilation is difficult to verify. Based on these issues, we cannot help but raise a question: is there some way to reduce the size of SC and increase the efficiency of air pollution control? To answer this question, we propose in this article a novel solar chimney with an inverted U-type cooling tower and a water spraying system (SCIUCTWSS). The scale of the SCIUCTWSS is almost the same as the experimental SCPP in Manzanares [3]. The differences are as follows: (1) an upside down U-shaped tower is used to replace the traditional chimney; (2) a water spraying system is installed at the turning point of the U-shaped tower, which will enhance the stack effect; and (3) a filtrating screen is placed near the entrance of the collector due to the low velocity of airflow in this position, and it is assumed to be helpful for the filtration process. The water spray method is utilized and it is very efficient in reducing $PM_{2.5}$ pollution. Moreover, it has excellent advantages such as rapidity, an already available technology, low cost, and a nature-like process [22]. The air which first enters the SC is filtered, then when it goes updraft and gets out at the top of the chimney where it is cooled down thanks to water evaporation, so that it can go downdraft, then the clean air can immediately improve the air quality in the spectrum of human activity.

5.2 Model Description

5.2.1 System Mechanism

Actually, an attractive approach that using a tower built as an inverted U-tube to implement such an expansion–compression cycle was reported by Oliver et al. [23]. From Fig. 5.1, the humid air expands and rises on the left side of the tower, when it past the top of the tower, cooling (evaporative cooling utilized in this article) is

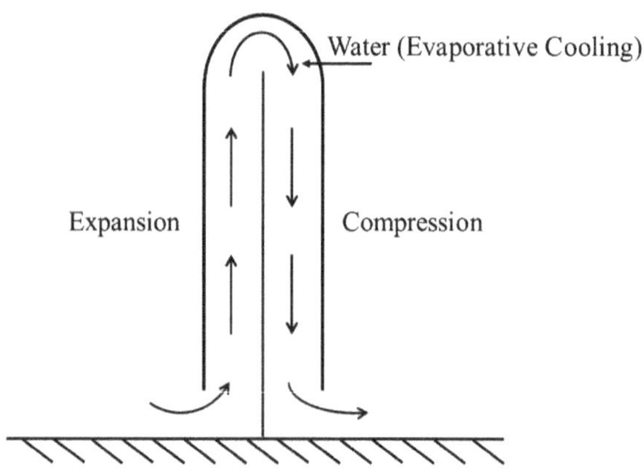

Fig. 5.1 Sketch of an inverted U-tube mechanism of expansion–compression cycle [23]

introduced. As a result, the air on the right side is cooler and denser than that of the left side. Thus, compression takes place as the air descends on the right side; an expansion-compression cycle could be realized via hydrostatics. In addition, they suggested that a large amount of power could be extracted during the expansion which represents the benefit. Conversely, a large but less amount of power is put back into the air during compression which represents the cost. The difference is the net output aiming to obtain. However, as they pointed, the scheme is only a concept due to it is not economically practical in construction of such a very high tower. In fact, this configuration is a form of a natural draft by the difference of hydrostatic pressure. If we can find some ways to augment this pressure difference, the cycle can still work as the tower is not very high. Benefit from the development of SC technology, which is able to produce a strong natural convection inside the tower and provide a promising approach. Based on this idea, we propose a solar chimney with an inverted U-type cooling tower and a water spraying system (SCIUCTWSS). Unlike the reported expansion-compression cycle, we do not need to extract output power in the expansion process due to the clear air through the processes of this system is the desired product.

5.2.2 Geometric Model

To investigate the effectiveness of the SCIUCTWSS, a simplified model is adopted for the numerical analysis. As shown in Fig. 5.2, the model has an inverted U-type tower with 200-m-height SC and a 200-m-height cooling tower; both of them have a radius of 5 m. There is a collector with a radius of 120 m and of 2 m height which covers the ground in a round shape. Since solar radiation heats the air inside the ground collector, airflow is driven by the buoyancy generated in the system and moves closer to the center of the collector where it goes upward inside the chimney due to the stack effect. Near the collector inlet there is a filtrating screen placed vertically with the total area being 1382 m^2 and the thickness being 2 m. When the airflow passes through the filtrating screen, $PM_{2.5}$ and large particulate matter are absorbed by the filtrating system and are removed from the air. Assuming the symmetric property to be perpendicular to the *z*-axis direction, only half of the whole system is displayed in the model. This assumption is acceptable with steady numerical simulation. The influence of energy storage layer was not considered and the geometrical model was not included. The position of water injection is at the top cooling tower. In this article, the main purpose is to study the performance of such a SC system with evaporative cooling, thus the energy required to pump the water to the top of the tower is not considered. In short, pushing volumes of water to the top cooling tower is not difficult. When the water is injected into the system, evaporative cooling occurs.

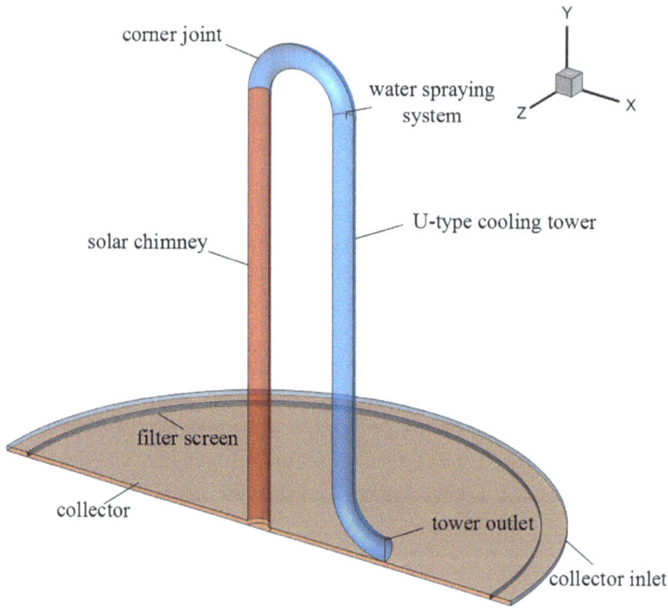

Fig. 5.2 3D geometrical model of the whole SC system

5.2.3 *Mathematical Model*

For a conventional SCPP, the airflow inside the system is considered to be natural convection induced by solar radiation heating the ground wall. Thus, the Rayleigh number is introduced to characterize the buoyancy-induced flow in the collector and the chimney:

$$Ra = \frac{g\beta \Delta T L^3}{a v} \tag{5.1}$$

where g is the gravitational acceleration, β is the thermal expansion coefficient, ΔT is the maximum temperature increase within the system, L is the collector height, α is the thermal diffusivity, v is the kinematic viscosity. The preliminary resulting values of Rayleigh number are higher than 10^{10} for the whole system. Therefore, the turbulent mathematical model needs to be selected to describe fluid flow within the system. The standard $k{-}\varepsilon$ turbulent model is chosen as an economic approach because of its robustness at a relatively low computational cost. The density variation of the air is caused by temperature changes, rather than that of the pressure. The incompressible flow is assumed and we use the ideal gas law to express the relationship between density and temperature for natural convection [24]. As a result, the transport equations for incompressible turbulent flow can be written as follows:

Continuity equation:

$$\frac{\partial \rho}{\partial t} + \frac{\partial}{\partial x_i}(\rho u_i) = S_{\mathrm{m}} \tag{5.2}$$

where the mass source term S_{m} is added to or removed from the continuous phase due to evaporation or condensation of the liquid droplets.

Navier–Stokes equation:

$$\frac{\partial}{\partial t}(\rho u_i) + \frac{\partial}{\partial x_j}(\rho u_i u_j) = -\frac{\partial p}{\partial x_i} + \frac{\partial \tau_{ij}}{\partial x_j} + \rho g_i + F_i \tag{5.3}$$

where the stress tensor τ_{ij} is defined as:

$$\tau_{ij} = \left[\mu\left(\frac{\partial u_i}{\partial x_j} + \frac{\partial u_j}{\partial x_i}\right)\right] - \frac{2}{3}\mu\frac{\partial u_i}{\partial x_i}\delta_{ij} \tag{5.4}$$

Energy equation:

$$\frac{\partial}{\partial t}(\rho E) + \frac{\partial}{\partial x_i}(u_i(\rho E + p)) = \frac{\partial}{\partial x_j}\left(k_{\mathrm{eff}}\frac{\partial T}{\partial x_j} - \sum_j h_j J_j + u_j(\tau_{ij})_{\mathrm{eff}}\right) + S_h \tag{5.5}$$

where k_{eff} is the effective conductivity ($k + k_{\mathrm{t}}$, where k_{t} is the turbulent thermal conductivity); J_j is the diffusion flux of species j; S_{h} includes the heat of chemical reaction or any other volumetric heat sources. $(\tau_{ij})_{\mathrm{eff}}$ is the deviatoric stress tensor, defined as:

$$(\tau_{ij})_{\mathrm{eff}} = \mu_{\mathrm{eff}}\left(\frac{\partial u_j}{\partial x_i} + \frac{\partial u_i}{\partial x_j}\right) - \frac{2}{3}\mu_{\mathrm{eff}}\frac{\partial u_k}{\partial x_k}\delta_{ij} \tag{5.6}$$

In the Eq. (5.5):

$$E = h - \frac{p}{\rho} + \frac{v^2}{2} \tag{5.7}$$

where sensible enthalpy h is defined for incompressible flows as:

$$h = \sum_j Y_j h_j + \frac{p}{\rho} \tag{5.8}$$

where Y_j is the mass fraction of species j.

Equation for the turbulent kinetic energy k:

$$\frac{\partial}{\partial t}(\rho k)+\frac{\partial}{\partial x_i}(\rho k u_i)=\frac{\partial}{\partial x_j}\left[\left(\mu+\frac{\mu_t}{\sigma_k}\right)\frac{\partial k}{\partial x_j}\right]+G_k+G_b-\rho\varepsilon-Y_M+S_k \quad (5.9)$$

Equation for the energy dissipation:

$$\frac{\partial}{\partial t}(\rho\varepsilon)+\frac{\partial}{\partial x_i}(\rho\varepsilon u_i)=\frac{\partial}{\partial x_j}\left[\left(\mu+\frac{\mu_t}{\sigma_\varepsilon}\right)\frac{\partial \varepsilon}{\partial x_j}\right]+C_{1\varepsilon}\frac{\varepsilon}{k}(G_{\mathrm{k}}+C_{3\varepsilon}G_b)-C_{2\varepsilon}\rho\frac{\varepsilon^2}{k}+S_\varepsilon \quad (5.10)$$

where G_{k} represents the generation of turbulence kinetic energy because of the mean velocity gradients and can be defined as $G_{\mathrm{k}}=-\rho\overline{u_i'u_j'}(\partial u_j/\partial x_i)$; G_{b} is the generation of turbulence kinetic energy due to buoyancy; σ_{k} and σ_ε are the turbulent Prandtl numbers for k and ε: $\sigma_{\mathrm{k}}=1.0$, $\sigma_\varepsilon=1.3$. $C_{1\varepsilon}$, $C_{2\varepsilon}$ are constants: $C_{1\varepsilon}=1.44$, $C_{2\varepsilon}=1.92$. $\mu_{\mathrm{t}}=\left(C_\mu\rho k^2/\varepsilon\right)$ and $C_\mu=0.09$.

Species transport equation:

$$\frac{\partial\rho Y_{\mathrm{H_2O}}}{\partial t}+\frac{\partial}{\partial x_j}(\rho Y_{\mathrm{H_2O}}u_i)=\frac{\partial}{\partial x_j}\left[\left(\rho D_{\mathrm{H_2O}}+\frac{\mu_t}{Sc_{\mathrm{t}}}\right)\frac{\partial Y_{\mathrm{H_2O}}}{\partial x_j}\right]+S_{\mathrm{H_2O}} \quad (5.11)$$

where $S_{\mathrm{H_2O}}$ is the water vapor added to or removed from the air because of evaporation or condensation.

Scalar quantities:

$$\frac{\partial}{\partial t}(\rho\phi)+\frac{\partial}{\partial x_j}(\rho\phi u_i)=\frac{\partial}{\partial x_j}\left(\Gamma\frac{\partial\phi}{\partial x_j}\right)+S_\phi \quad (5.12)$$

where ϕ is an arbitrary scalar, Γ is a diffusion coefficient.

5.2.4 Boundary Conditions

The boundary conditions of the computations are shown in Table 5.1.

In this article, the airflow is assumed fully developed and the ambient air temperature constantly at 293 K. The ambient relative humidity is set as 58.5%. Relative static pressure is used for the simulation to analyze the whole pressure distribution of the system, which is the static pressure difference between the SCIUCTWSS and the environment at the same height (set as 0 in this article) [25], also used by Pastohr et al. [6], Ming et al. [26], and Sangi et al. [27]. Convection heat transfer occurs between the canopy of the collector and the ambient air, the heat transfer coefficient is set as 8 W/(m^2 K) [13] which can be accepted when the ambient air velocity is not very large. The solar radiation is set at 857 W/m^2, so

Table 5.1 Boundary conditions

Place	Boundary type	Value
Collector inlet	Pressure inlet	p = 0 Pa, T = 293 K, RH = 58.5%
Chimney outlet	Pressure outlet	p = 0 Pa
Surface of the chimney	Wall	$q = 0\ W/m^2$
Surface of the tower	Wall	$q = 0\ W/m^2$
Filtrating screen	Fan	0 Pa
Ground wall under the canopy	Heat flux	$600\ W/m^2$
Collector canopy	Wall	T = 293 K, $h = 8\ W/(m^2\ K)$
Chimney and tower surface	Adiabatic wall	$0\ W/m^2$
Symmetry surface	Symmetry	

the corresponding heat flux on the ground surface is set at 600 W/m^2 due to the energy loss through thermal radiation and conduction [28]. As the chimney is not too high, the pressure at both the entrance of the collector and the exit of the chimney are set equal to the standard atmospheric pressure. The filtrating screen adopts fan boundary condition to simulate the pressure drop. The pressure drop across the filtrating screen can approximate it by Lackner et al. [29]:

$$\Delta p = \rho v^2 \tag{5.13}$$

Thus, the pressure drop is much small that can be neglected.

5.2.5 Numerical Method

The governing equations are solved by the finite volume method in the general purpose CFD program ANSYS Fluent. SIMPLE algorithm is applied as the pressure-velocity coupling scheme. The QUICK scheme is used to discretize the convective terms and the second order upwind scheme is chosen as the spatial discretization method for the diffusion terms. Besides, the numerical calculation is performed with the double precision solver due to the disparate length scales in the model. To monitor the solution convergence, the iterations were continued until the relative errors for all variables were below 10^{-4}.

To verify the grid-independent performance of the numerical simulation results, three test cases of the model under the same conditions (solar radiation is 857 W/m^2, injected water is 0 kg/s) with grid numbers being 1,989,315; 2,344,117; 2,665,165 were verified. Numerical simulation results showed that the volume flow rates of the system outlet are 105.76; 106.81; 106.82 m^3/s, and the average temperature of the system outlet are 339.69; 339.78; 339.45 K, respectively. From the comparison between the numerical simulation results, we found that there was only a deviation of approximately 1.0% between these three results, which demonstrated the solutions in this article are grid-independent. The grid number of 2,344,117 is selected in this paper.

5.3 Results and Analysis

When the solar radiation through the transparent canopy of the collector is absorbed by the ground, the ground temperature rises, then heating the air inside the collector. Resulting in the density and the relative humidity of air reduced, strong updrafts of natural convection are formed by air buoyancy difference. When the air flows to the top of the cooling tower, liquid water is injected equally and uniformly along the injection surface. The diameter and the temperature of the water liquid are assumed to be 30×10^{-6} m and 280 K, respectively. Because the air is dry, liquid water once injected, evaporative cooling would occur, then the air becomes heavier so that a downdraft is formed.

To determine the effectiveness of air pollution mitigation, the flow performances of the SC system is investigated. Studying the relative static pressure changes within the system which is the cause of the driving force is necessary. Besides, the temperature characteristics of the system play a crucial role for the natural flow. Air changes in properties caused by temperature variation also provided the basis to measure the flow performances, such as the air density and the air relative humidity distribution in the system.

Figure 5.3 displays the static pressure when the amount of injected water differs from each other. From this figure, it is obvious that different amounts of injected water affect the relative static pressure distribution of the SC system. A similarity between these figures is that the maximum pressure appears at the top of the U-shaped structure. The minimum pressure always presents at the bottom of the chimney, and then the relative pressure increases gradually through the chimney. These two can be attributed to the updraft.

Once liquid water is injected, part of the airflow flows downward, so the value of the maximum pressure decreases with the increasing injected liquid water, as shown in Fig. 5.3a–d. Besides, as the solar radiation heating the air inside the collector, thermal airflow is generated and vertical natural convection formed. Because of the stack effect, strong vertical natural convection is produced inside the chimney. Correspondingly, it means that there is greater buoyancy, therefore the negative pressure or the minimum pressure represents the pressure difference between the airflow within the system and the stable atmosphere outside. Ambient air is sucked into the bottom of the chimney. In some ways, the value of the minimum pressure can reflect the strength of natural convection [25]. The greater the negative pressure, the stronger the natural flow. By contrasting the relative static pressure distributions as shown in Fig. 5.3a–d, we can find that the minimum pressure at the chimney bottom are −13.50, −68.50, −105.00, −136.00 Pa, respectively corresponding to the injected water: $q = 0$ kg/s, $q = 1$ kg/s, $q = 2$ kg/s, $q = 3$ kg/s. It is evident that the natural flow undergoes a promotion by liquid water injected in the system.

Figure 5.4 shows the comparison of contours of velocity distributions at the symmetry plane when the injected water increased from 0 to 3 kg/s. It is apparent that when no liquid water is injected into the system, as shown in Fig. 5.4a, the natural flow inside the system is not strong. If no measures are taken to make the air

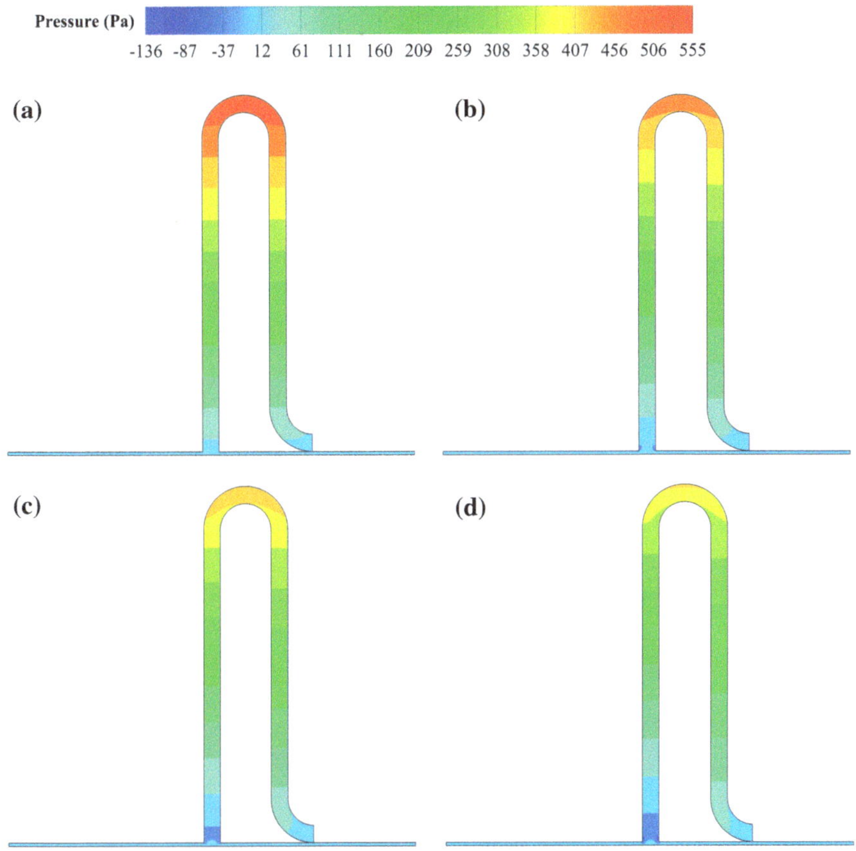

Fig. 5.3 Influence of injected water on relative static pressure distributions in the symmetry plane. **a** q = 0 kg/s, **b** q = 1 kg/s, **c** q = 2 kg/s and **d** q = 3 kg/s

flow to go downward, the volume flow rate of airflow moved to the outlet is low. Meanwhile, the reversed flow phenomenon is found near the system outlet. This means that the flow inside the system is very weak at this time, needed to take measures to strengthen the natural flow. By water injection in the second part of the inverted U shaped chimney, the velocity of airflow increases with the increasing amount of injected water. As air nearby flows into the collector, heated by the solar radiation and is sucked into the chimney because of the negative pressure. The updraft reaches its peak speed from 9.50 to 13.50 m/s at the chimney bottom, as shown in Fig. 5.4b–d. Velocity distributions within the system are very similar, even at the very place where the most intense flow shows up. The only difference is that the magnitude of the velocity is gradually increasing. The increase of velocity is due to the absorption of liquid water by the airflow, which also becomes heavier. The driving force represented by the negative pressure is getting stronger. Strong natural flow formed inside the system. What is noteworthy is that the reversed flow

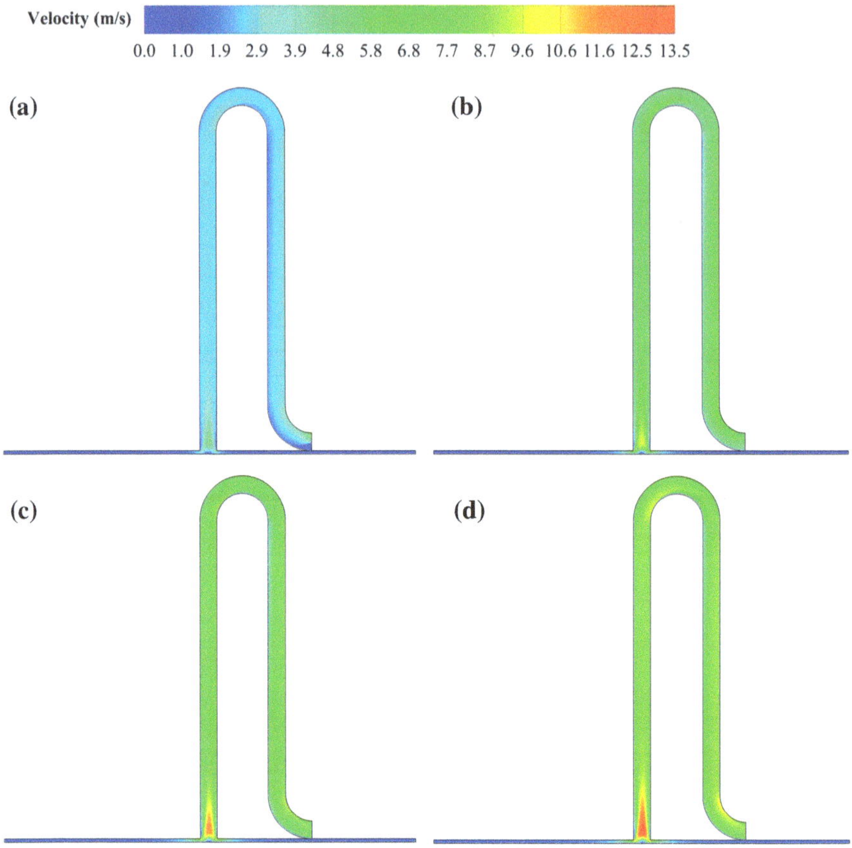

Fig. 5.4 Influence of injected water on velocity distributions in the symmetry plane. **a** q = 0 kg/s, **b** q = 1 kg/s, **c** q = 2 kg/s and **d** q = 3 kg/s

phenomenon disappears in these three figures, which provides an additional evidence for the strong natural flow. In addition, we found that when air flows through the U-turn joint, there will be some flow losses. It indicates that wall friction mainly contributes to the depletion of kinetic energy of airflow as the air flows inside the system.

In order to further illustrate the flow performances in the system, Fig. 5.5 shows the comparison of contours of velocity distributions inside the collector. As it is seen, airflow nearby is sucked into the chimney, and the closer the chimney, the greater the speed. Therefore, the structured mesh in this area is considered small enough to have acceptable results. At the very center of the collector, there is a small area of stagnation, where the magnitude of speed is relatively small. It is because of the airflow expansion. Moreover, as the water into the airflow in the cooling tower, the velocity distribution in the collector also happened to change. With the increase of water injection, the natural flow near the chimney is getting

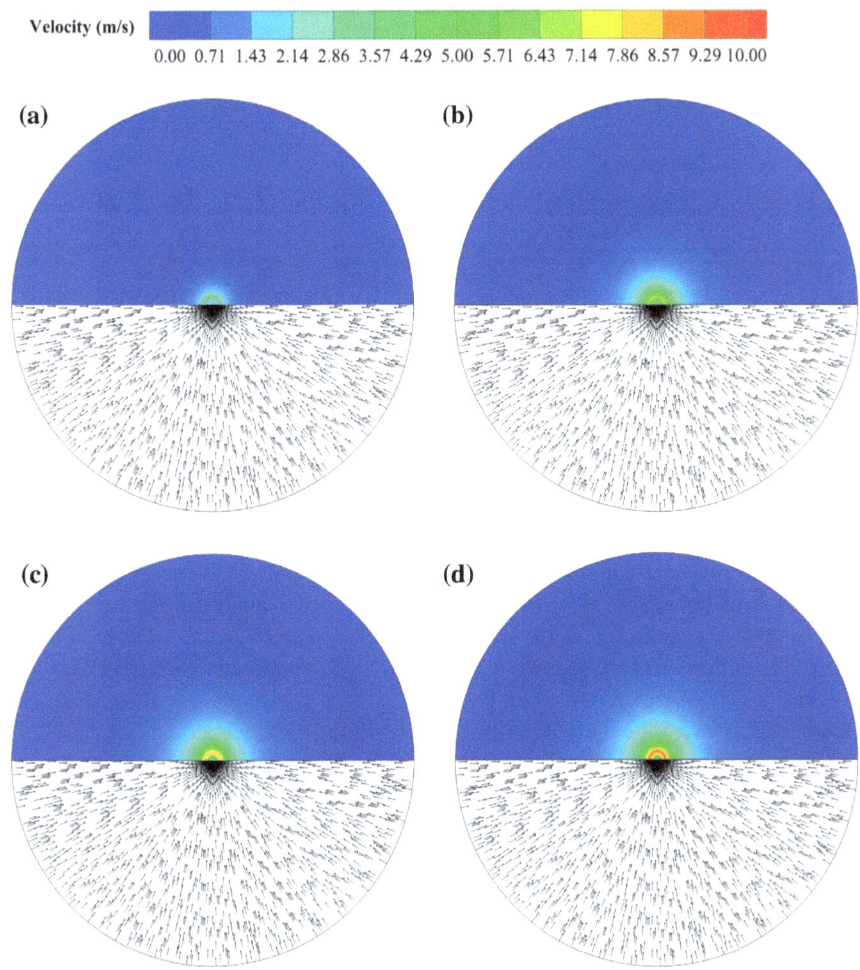

Fig. 5.5 Influence of injected water on velocity distributions in the collector cross section ($y = -1$ m). **a** $q = 0$ kg/s, **b** $q = 1$ kg/s, **c** $q = 2$ kg/s and **d** $q = 3$ kg/s

stronger, as shown in Fig. 5.5a–d. This shows again that water injection is useful which can strengthen the natural convection, whether it is for SC, cooling tower or collector. Whereas, the influence of ambient crosswind is not taken into account in this paper, future studies should consider this influential factor.

Figure 5.6 denotes the influence of injected water on the average velocity and volume flow rate of airflow. It is evident that with the increasing amount of injected water, the velocity of airflow increases not only in the collector but also in the chimney. In fact, the natural flow throughout the system has been strengthened, as shown in Fig. 5.6. As for the reason that there is little variation of velocity in the collector inlet, it is mainly because of the large radius of the collector, thus the effect

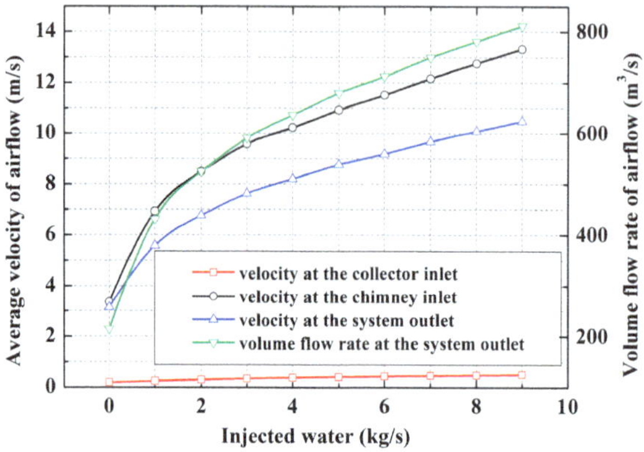

Fig. 5.6 Influence of injected water on the average velocity and volume flow rate of airflow

on the flow at this place is relatively small. But even so, we can still see a slight growth trend from the curve of velocity at the collector inlet. When thermal airflow is sucked into the chimney, reaches a maximum speed at the chimney inlet or the chimney bottom, which is consistent with our above analysis. However, it is apparent that the velocity at the system outlet is lower than that of the chimney outlet. With the increase of water injection, this comparative decrease is also growing. It is because on one hand, with the increasing velocity, the flow losses are become larger when the thermal air flows through the chimney, corner joint and the cooling tower also increased. On the other hand, when thermal air rises to the top of the system, since it will not immediately be cooled, it will gradually accumulate at the top so that the local pressure is relatively large. To solve this problem, it should be found a way to make it cool instantly and drop. Therefore, evaporating cooling is not the only method of cooling; the flow performance of the system is expected to improve in future studies. But even so, from the figure, the volume flow rate of airflow in the system is remarkable. If the amount of injected water is 9 kg/s, this system is able to process atmospheric air at a volume flow rate of 810 m^3/s, corresponding to the volume of 69,984,000 m^3 (nearly 0.07 km^3) of air to be cleaned in one day. Of course, the efficiency of the performance and the economy should also be taken into comprehensive consideration into an actual situation.

5.4 Conclusion

A novel solar chimney with an inverted U-type cooling tower and a water spraying system (SCIUCTWSS) is proposed to mitigate urban air pollution. A filtrating screen is placed near the entrance of the collector, $PM_{2.5}$ and large particulate matter is removed from the air in the natural flow. The water spray method is utilized at the top of cooling tower for evaporative cooling, it causes the air to become heavier and a natural air downdraft to be produced, then the clean air out of the system outlet can immediately improve the air quality in the spectrum of human activity. Water injection is efficient to strengthen the natural convection, whether it is for SC, cooling tower or collector. If the amount of injected water is 9 kg/s, this system is able to process atmospheric air at a volume flow rate of 810 m^3/s, corresponding to the volume of 69,984,000 m^3 of air to be cleaned in one day.

References

1. J.r. Schlaich, R. Bergermann, W. Schiel, G. Weinrebe, Design of commercial solar updraft tower systems—utilization of solar induced convective flows for power generation, Journal of Solar Energy Engineering, 127 (2005) 117–124.
2. W. Haaf, K. Friedrich, G. Mayr, J. Schlaich, Solar Chimneys Part I: Principle and Construction of the Pilot Plant in Manzanares, International Journal of Solar Energy, 2 (1983) 3–20.
3. W. Haaf, Solar chimneys: part ii: preliminary test results from the Manzanares pilot plant, International Journal of Sustainable Energy, 2 (1984) 141–161.
4. N. Pasumarthi, S. Sherif, Experimental and theoretical performance of a demonstration solar chimney model—Part I: mathematical model development, International Journal of Energy Research, 22 (1998) 277–288.
5. N. Pasumarthi, S. Sherif, Experimental and theoretical performance of a demonstration solar chimney model—Part II: experimental and theoretical results and economic analysis, International Journal of Energy Research, 22 (1998) 443–461.
6. H. Pastohr, O. Kornadt, K. Gürlebeck, Numerical and analytical calculations of the temperature and flow field in the upwind power plant, International Journal of Energy Research, 28 (2004) 495–510.
7. C.B. Maia, A.G. Ferreira, R.M. Valle, M.F. Cortez, Theoretical evaluation of the influence of geometric parameters and materials on the behavior of the airflow in a solar chimney, Computers & Fluids, 38 (2009) 625–636.
8. S.K. Patel, D. Prasad, M.R. Ahmed, Computational studies on the effect of geometric parameters on the performance of a solar chimney power plant, Energy Conversion and Management, 77 (2014) 424–431.
9. A. Koonsrisuk, T. Chitsomboon, A single dimensionless variable for solar chimney power plant modeling, Solar Energy, 83 (2009) 2136–2143.
10. H.F. Fasel, F. Meng, E. Shams, A. Gross, CFD analysis for solar chimney power plants, Solar energy, 98 (2013) 12–22.
11. W.B. Krätzig, Physics, computer simulation and optimization of thermo-fluidmechanical processes of solar updraft power plants, Solar Energy, 98 (2013) 2–11.
12. F. Cao, H. Li, L. Zhao, T. Bao, L. Guo, Design and simulation of the solar chimney power plants with TRNSYS, Solar Energy, 98 (2013) 23–33.
13. M.d.S. Bernardes, T. von Backström, D. Kröger, Critical evaluation of heat transfer coefficients applicable to solar chimney power plant collectors, in: Proceedings of ISES World Congress 2007 (Vol. I–Vol. V), Springer, 2008, pp. 1706–1713.
14. A. Koonsrisuk, Mathematical modeling of sloped solar chimney power plants, Energy, 47 (2012) 582–589.
15. F. Cao, L. Zhao, L. Guo, Simulation of a sloped solar chimney power plant in Lanzhou, Energy conversion and management, 52 (2011) 2360–2366.
16. S. Kalash, W. Naimeh, S. Ajib, Experimental investigation of the solar collector temperature field of a sloped solar updraft power plant prototype, Solar Energy, 98 (2013) 70–77.
17. B. Kashiwa, C.B. Kashiwa, The solar cyclone: A solar chimney for harvesting atmospheric water, Energy, 33 (2008) 331–339.

18. T. Ming, T. Gong, R.K. de Richter, W. Liu, A. Koonsrisuk, Freshwater generation from a solar chimney power plant, Energy Conversion and Management, 113 (2016) 189–200.
19. Q. Cao, D.Y. Pui, W. Lipiński, A Concept of a Novel Solar-Assisted Large-Scale Cleaning System (SALSCS) for Urban Air Remediation, Aerosol and Air Quality Research, 15 (2015) 1–10.
20. X. Zhou, Y. Xu, S. Yuan, C. Wu, H. Zhang, Performance and potential of solar updraft tower used as an effective measure to alleviate Chinese urban haze problem, Renewable and Sustainable Energy Reviews, 51 (2015) 1499–1508.
21. T. Ming, W. Liu, S. Caillol, Fighting global warming by climate engineering: Is the Earth radiation management and the solar radiation management any option for fighting climate change?, Renewable and Sustainable Energy Reviews, 31 (2014) 792–834.
22. S. Yu, Water spray geoengineering to clean air pollution for mitigating haze in China's cities, Environmental chemistry letters, 12 (2014) 109–116.
23. T.K. Oliver, W.N. Groves, C.L. Gruber, A. Cheung, Energy from humid air. Final report, 1979, in, http://www.osti.gov/scitech/servlets/purl/5715593/. Doi:10.2172/5715593.
24. Flunet Inc, Fluent user's guide, in, 2006.
25. M. Tingzhen, L. Wei, X. Guoliang, Analytical and numerical investigation of the solar chimney power plant systems, International Journal of Energy Research, 30 (2006) 861–873.
26. T. Ming, R.K. Richter, F. Meng, Y. Pan, W. Liu, Chimney shape numerical study for solar chimney power generating systems, International Journal of Energy Research, 37 (2013) 310–322.
27. R. Sangi, M. Amidpour, B. Hosseinizadeh, Modeling and numerical simulation of solar chimney power plants, Solar Energy, 85 (2011) 829–838.
28. T. Ming, X. Wang, R.K. de Richter, W. Liu, T. Wu, Y. Pan, Numerical analysis on the influence of ambient crosswind on the performance of solar updraft power plant system, Renewable and Sustainable Energy Reviews, 16 (2012) 5567–5583.
29. K.S. Lackner, S. Brennan, J.M. Matter, A.-H.A. Park, A. Wright, B. Van Der Zwaan, The urgency of the development of CO2 capture from ambient air, Proceedings of the National Academy of Sciences, 109 (2012) 13156–13162.

Index

T. Ming et al., *Pollutant Dispersion in Built Environment*,
DOI 10.1007/978-981-10-3821-1